Who Will Win The Big Game?

50 Championship Characteristics

A Psychological
And
Mathematical Method
For Identifying
Winning Players, Teams
And Coaches

Jay P. Granat, Ph.D.

&

Carlton J. Chin, CFA

Published by World Audience, Inc.
(www.worldaudience.org)
303 Park Avenue South, Suite 1440
New York, NY 10010-3657
Phone (646) 620-7406; Fax (646) 620-7406
info@worldaudience.org
Edited by M. Stefan Strozier
ISBN 978-1-935444-37-4

$20
© 2010, Jay P. Granat, Ph.D. and Carlton Chin, CFA

World Audience (www.worldaudience.org) is a global consortium of
artists and writers, producing quality books and the literary journal
audience, and *The audience Review*. Our periodicals and books are
edited by M. Stefan Strozier and assistant editors. Please submit
your stories, poems, paintings, photography, or artwork:
submissions@worldaudience.org.

Inquire about being a theater reviewer: theatre@worldaudience.org.
Thank you.

Who Will Win The Big Game?

50 Championship Characteristics

by

Jay P. Granat, Ph.D.

&

Carlton Chin, CFA

January, 2010

A World Audience Book

New York, NY (USA)
Newcastle, New South Wales (Australia)

- Two great players are meeting in a championship. Who will win and why?
- Two top undefeated teams are meeting in the big game. Which one is most likely to win the big contest?
- Why do some athletes excel in the big game?
- What causes others to choke when the pressure is on?
- What role does psychology play in determining the outcome of sporting events?
- Can psychological factors determine how well an athlete or a team performs on a particular day?
- What are the key psychological, sociological and emotional variables that impact the way teams, athletes and coaches perform under pressure?
- What causes losing streaks and what causes winning streaks?
- Why can some coaches win with many different teams and players?
- What allows some athletes to play in the zone when the pressure is on?
- What causes some athletes to choke and perform poorly in big games?
- Why do some athletes seem to always win while other equally talented players seem to always lose in the big game?
- What percentage of sports is mental?
- What are some of the secrets to winning the big game?
- What can math and statistics tell us about who will win the big game?
- Do statistics really support the importance of psychological factors in determining the winners and losers of big games?
- What should sports executives, coaches, managers, players and athletic directors understand about sport psychology and sports statistics?
- What are the most crucial statistics for forecasting the winners of the big game?

To all the wonderful athletes, coaches and parents I have had the pleasure of working with for many years. You have taught me a lot about what it takes to win the big game.

Jay P. Granat, Ph.D.

To everyone who strives to improve focus and performance. In life, there are certain key moments when you want to increase your intensity and "bring it." I hope that our work will help you make the most of these moments in life.

For my friends and family, thank you for your input, friendship, and support through the years.

And especially for R and JR, and those closest to me, who encourage me to do what I love. Thank you.

Carlton J. Chin

What Makes This Book Different?

There are lots of books on sport psychology and there are many guides which present information about sports and statistics. Carlton and I thought that a book which combined the disciplines of psychology and mathematics would be quite helpful and quite interesting to a wide range of people.

We are already starting to see how an approach which combines these two disciplines can be very valuable to athletes, coaches, parents of athletes, teams, leagues, athletic directors, team owners and sports executives.

The ideas in this book can be helpful in recruiting the right athletes, developing players, improving performance and in building teams. Psychological factors impact statistics – and statistics have an effect on behavior. Solving performance problems with information from both disciplines will empower players, coaches, sports executives, athletic directors and teams.

We are currently developing models which will be valuable for people involved in all of the major sports. Some of our initial ideas and findings are presented in this book. We will also identify promising areas for additional research.

As you will see, some of the factors which impact the outcome of the big game are easier to quantify than others. We believe that hard statistical data as well as observations of a lot of athletes in a lot of big games can yield some fascinating information as to what distinguishes the winners from the losers in these big events.

The Big Game

Events like The Super Bowl, The World Series, A Championship Fight, The Masters, Wimbledon, The US Open, The Stanley Cup, The NCAA Basketball Championship, The World Cup, The Olympic Finals, The BCS Championship, a State Championship, or a Club Championship, evoke special interest and a wide range of emotions for both the fans and the participants.

Some people collapse under the pressure and others thrive in the spotlight and relish these big moments.

What determines which teams, athletes and coaches will succeed when they are in the big game? And what determines who will fail to perform to their potential on the big stage?

As was mentioned earlier, this book will address the psychological forces which contribute to winning and to losing these all-important contests. Whenever possible, this book will also present relevant mathematical data to help readers understand what really affects the outcomes in these major sporting events.

Who Should Read This Book?

The ideas and information in this book will be of interest to a wide range of sports enthusiasts.

If you are an owner of a team, an executive for a team, an athletic director or a manager who wants your team to perform better in the big game, this book should be quite valuable for you and for your organization.

If you are a professional athlete, high school athlete, college athlete or a weekend warrior, you can get a better understanding of sport psychology and sports statistics from this book.

If you are a coach who wants your team to play better, this guide should give you some new ways to think about your team and some new ideas about how to manage and how to motivate your players.

If you're a journalist who writes about sports, you will get some increased understanding about the psychology and about the statistics of winning and losing the big game.

If you're a parent who wants your child to master the mental aspects of competing in sports, this book will contain some useful information for you.

If you are a coach or a recruiter who is trying to decide which athlete should join your team, you should find the ideas in this book to be helpful in making some of your personnel decisions.

If you like to debate the outcome of sporting events with your friends, co-workers or colleagues, you will get some additional fuel to support your viewpoint from this book.

If you're a sports fanatic who likes to call in to sports talk radio shows to express your opinion on big games, this book will give you some new ways of thinking about the teams you love and the teams you hate.

If you are a manager, a business owner, a coach or a CEO who wants to learn more about what helps people to perform their best, you may get some useful tips and insights from the pages that follow.

If you like to bet on sporting events, this book will give you some additional insights into the psychology and statistics of competing effectively in individual and in team sports.

While this book is not intended to encourage gambling, millions of Americans do wager on sporting events every day. Fantasy leagues are now quite popular among baseball and football fans. And events like The Super Bowl and The Final Four attract lots of dollars from casual fans and from serious gamblers.

The ideas and information in this guide can be incorporated into existing models and systems which people use to handicap sporting events.

Last, readers of this book can utilize the information to evaluate and to compare one competitor against another with regard to mental toughness.

Quant Fact: Project Evolution And A Note From Carlton

This project is the result of "ideas," which Jay and I have bounced around over the years, "numbers" that my friends and associates know that I love to play with – and on a personal level, the culmination of a long-time friendship. I enjoy working with my good friend, Jay – and have especially enjoyed seeing how the dynamics of team chemistry can increase energy levels exponentially. We hope that others will enjoy our research – and if nothing else, have enjoyed the "exclamation point" that our work has put on our friendship.

Scorecards To Help You Evaluate Teams, Coaches And Players Prior To The Big Game

At the end of this book, you will find two scorecards. One includes all of the variables mentioned in this book. As was mentioned earlier, some of these factors are easier to quantify than are others. Carlton and I will continue to study many of these other factors in the near future, so that we might learn more about the elements which create success in the big game.

The second scorecard is based on hard data described in the "quant facts" which appear throughout this book.

Use one scorecard or both and see which approach you like the most and determine which one works best for you and for the kind analysis you are performing for your sport.

As I noted earlier, you may want to add our ideas and methods to the current system or model you are utilizing to do your own analysis.

Sports, Psychology And Mathematics

There is much debate about what percentage of sports is psychological. Some might say that golf at the professional level is ninety-five per cent mental. Others might say that psychology plays a significant role in football or baseball games. However, very few people would say that psychology has no impact on the outcome of big games in virtually every sport.

This book will highlight some of the key factors related to sport psychology which impact the outcome of championship contests and other big games.

In addition, it will introduce some statistics which substantiate the role which some of these mental factors play in determining the winner and the loser of major sporting events.

Table of Contents

Introduction

I have counseled thousands of athletes, coaches and parents of athletes for quite some time. In fact, every week I get calls from athletes, parents of athletes and coaches who are struggling. When they call, they are often frustrated, losing and lacking in confidence and focus. They are afraid of losing, stuck in a slump and their motivation may be waning. And in many instances, they are concerned about an upcoming big game.

My clients have included Olympic champions, professional golfers, professional tennis players, professional baseball players, jockeys, martial artists, gymnasts, boxers, divers, swimmers, hockey players, skiers, pool players, basketball players, ice skaters, football players, wrestlers, lacrosse players, motorcycle racers, cyclists, weight lifters, pole vaulters, soccer players, hockey players, shooters, long distance runners, cowboys and cowgirls.

Consequently, I have learned a great deal about the psychological factors which impact the minds of individual athletes and teams. In addition, I believe I have acquired a good understanding of how winners think and behave and how losers think and how they behave.

Since psychological factors do not exist in a vacuum, it is important to get a sense as to how some key variables might interact with one another and how key statistics reflect the psychological strengths and weaknesses of competitors.

For example, an athlete who is skilled at distraction control, might be less prone to make a mental error in the big game. As you will see, things like turnovers and unforced errors play a big role in determining the outcomes of big games.

The ideas presented in this book have merit on their own. However, in order to do a thorough analysis, some readers will want to combine our ideas along with other traditional sports data when trying to identify the like winner of big games.

For instance, when evaluating baseball teams, you must consider factors like earned run average, batting average, winning streaks, losing streaks, home record, away record, pitch count, time of the season, lefty-righty match ups, runs allowed, margins of victory, margins of losses, records against winning teams, records against losing teams, team speed, stolen bases and walks allowed along with the variables cited in this guide.

Likewise, when analyzing football teams, you need to take into account the quality of the quarterback, points allowed, points scored, home field, previous history, weather conditions, road record, injuries, time of possession as well as the mental toughness and cohesiveness of the competing teams.

When comparing basketball teams, you can not overlook the number of rebounds, the shooting percentage, foul shooting percentage, the size of the front line, the number of assists that the point guards get each game, winning streaks, losing streaks, three point shots made, and turnovers as well as the teams' psychological strengths and weaknesses.

Who Will Win The Big Game?: A Psychological And Mathematical Method For Identifying Winning Athletes, Teams And Coaches will delineate some of the major variables that tend to produce victories and create losses during the regular season and in the big game.

Readers will begin to understand some of the relationships between psychology, mental toughness, peak performance, mathematics, probability and traditional sports statistics.

Quant Fact: Our "Scientific" Approach

We can use widely-available statistics to help us predict winners. However, in order to capture information related to performance, we try to avoid "general" statistics and **attempt to carve out information by focusing on data specific to sports psychology and winning factors**. We focus on factors related to mental toughness, leadership, experience, and "successful, focused, execution" (such as consistent and high-level play, with minimal errors).

We examine how much knowledge can be gleaned from sports results and statistics. A variety of analytical tools were used to study the outcomes and historical data surrounding major sporting events. In particular, we focused on key psychologically-related factors that might be useful in predicting the outcome. We applied statistical factor analysis, regression analysis and correlations – as well as other mathematical models – across the major US sports. **Overall, we're pleased that historical sports data** *verifies many of the concepts of sport psychology, producing some interesting – and sometimes surprising – results.*

Individual Sports And Team Sports

Obviously, it is easier to evaluate competitors in an individual sport than it is to judge the large group of athletes who participate in some team sports. Not surprisingly, there are many more issues to consider when looking at a twenty five or fifty athletes as opposed to one individual competitor.

So, predicting the winner of a boxing contest, a tennis match or a two person golf match, is significantly easier than it is to predict the winner of a baseball game or of a football contest.

Nevertheless, we feel we have discovered some fascinating information as to what really allows a team or an individual to prevail in a big contest.

Who Will Win?
Factors That Impact
The Big Game

Confidence

Believing in one self and in one's ability is a key to performing well and to winning. As I have told many athletes, "It all begins with believing in your self and in your ability."

In fact, it is impossible to perform well without being self-confident.

Winning athletes and winning teams are confident, but they are not grandiose or arrogant. They find the right level of confidence and maintain a respect for the opponents.

Sometimes, athletes have an inflated opinion about themselves and their skills. In some instances, these athletes are masking some insecure feelings with their bravado.

I do not believe that these kinds of competitors do as well as those who believe in themselves, but who are not "over the top" in their self-appraisal.

In evaluating athletes or teams, favor competitors who exude a sincere and honest belief in themselves.

However, given a choice between an athlete or team who lacks confidence and who is somewhat overconfident, I would choose the competitor who appears to believe in himself or herself. In my experience, athletes who are acting or pretending to confident often start to believe and behave in a confident manner. This is desirable and you would rather have athletes err on the side of confidence as opposed to a lack of confidence.

A competitor or a team that appears anxious, nervous or self-doubting is not likely to perform well in the big game.

Thinking positively, believing in yourself and trusting your ability, skills and talent rarely hurt one's performance in a big game.

Previous Experience
In The Big Game

I believe that a team or player who has been to "the big stage" before can have an advantage over an opponent who is playing at this kind of venue for the first time.

Being in a major event helps coaches and players to get comfortable with the major event and helps them to manage the thoughts, feelings, distractions and experiences which accompany this kind of contest.

So, favoring a team that has been down this road before seems to make a great deal of sense. Our data below would seem to strongly support this concept.

What Do the Numbers Say?
Quant Fact:
"Big Game Experience" and the NBA

"Big Game Experience" and confidence are somewhat related, so this is a good opportunity to introduce and quantify one of the biggest factors in determining "who will win the big game." As we might expect, "experience" helps, but it is a stronger factor and more pervasive than we imagined.

Our analytical research of NBA Finals shows that experience and coaching are major factors in winning championships. Experience has been a common theme in the results of our research, across all sports. And there is little wonder to this finding. Experience means that you have been there before – so there are fewer butterflies. Teams and players can better focus on the task at hand: winning!

We give a team credit for "experience" if they played in a Championship Final over the past three years. *Using this simple method to measure "team experience" in the NBA Finals has been able to predict 9 of 12 winners, going 9-3 (75%).* This is based on twenty years of NBA Finals data, from 1990-2009.

Chart: "Big Game Experience" as Factor (Championship Winning Percentage in Recent Years for Teams with "More Experience")

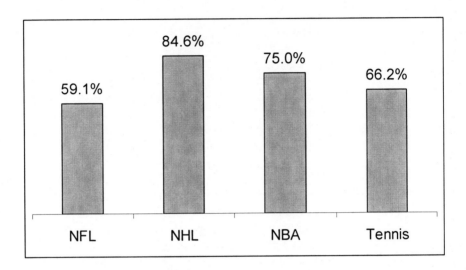

Results have been consistently positive across all sports as you can see in the chart. Although other factors and inter-relationships are involved (such as truly great teams getting to the finals multiple times), few would argue that experience helps your confidence – and both experience and confidence – are good things to have on your side!

Here's an additional tidbit about the NBA which leads into the next variable discussed in this book: Over this twenty-year time period, great coaches such as Phil Jackson and Gregg Popovich put together a combined 14-2 record in the NBA Finals. In the past, all-time great Red Auerbach led his Boston Celtics to a 9-2 record in NBA Finals. Luck? We don't think so.

And you will learn more about the roles of coaching in the pages that follow.

Coaches, Coaching And Leadership

Great coaching has a huge impact on confidence and on the development of an athlete. Some athletes report having one wonderful coach and others recall having several coaches who made significant impressions on them.

Top coaches are surrogate parents, motivators, teachers, confidants, team builders, strategists, talent scouts, generals and team leaders. These leaders know when to be tough on a player and when to be gentle with an athlete.

In fact, teams often can take on their coaches character as they integrate their leader's philosophy and approach to their game into their lives and their personas.

Great coaches have the ability to win with different kinds of teams with different kinds of players and in different kinds of situations. Great coaches are great psychologists. They have the capacity to inspire their players, build team chemistry and they can empower underdogs to beat teams that appear to be more powerful.

When comparing one team to another or one player to another, pay very close attention to the quality, experience, ability and track record of the respective coaches.

John Wooden, Phil Jackson, Pete Carroll, Pat Summitt, Joe Paterno, Mike Krzyzewski and Joe Torre are fine examples of coaches who do a lot of things very well. These coaches know how to prepare players for the big game on a consistent basis.

Quant Fact: Measuring the Difference a Coach Can Make

Can a coach really make that much of a difference? Some would argue that:

- *"It's just luck. There are bound to be some coaches that are more successful than others."* Or,

- *"The players are the ones winning the championships. That team would have won with me coaching them!"*

While we agree that there is some validity to these statements, we ***do*** believe that good coaches add value. Teams need leaders on the field, but they can also benefit from additional leadership on the bench. Although coaches are working constantly, we believe that the impact of coaches and managers can be particularly strong at crucial moments like championship series.

An interesting way to value a team's coach or manager is the team's performance when they are in that ultimate position to win a championship title. Great coaches win an inordinate amount of championships. When their teams are in a position to bring home the big prize, the coach helps the team to "bring it" and leads them to victory. Why do we think this is true?

Quant Fact: Studying Historical Results for Coaches

We studied the results of coaches – across multiple sports – who have guided their teams to multiple title shots. We then compared these results to our "lab experiment" and modeled teams and coaches with a random expectation of winning. We assumed that at this level of play (that is, when two teams are facing off for a championship), they each have a 50/50 chance of winning the title.

We compared the results of the real-life coaches and their multiple title shots – to our mathematical model. For the purposes of this study, we used a sampling of coaches who have reached at least three or four (or more, depending on the sport) championship games or series in their sport. For college football, we used Bowl Games with National Title implications.

Charts: Coaches and Championship Games/ Series Expected and Actual Winning Percentage

Actual Performance in Multiple Championship Appearances Is Better Than Expected, Implying "Skill"

(Normally 4 or More Appearances)

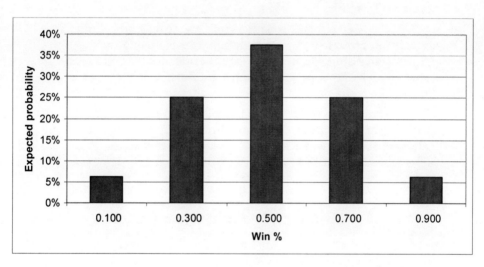

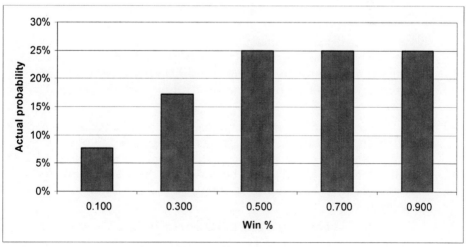

Numbers Don't Lie
Quant Fact: Great Coaches *DO* Exist

As you can see from the graphs, the real-life coaches have won many more titles than expected. In fact, there are many coaches who have won 80% or more of their title shots. This is a much higher winning percentage than expected. In addition there are fewer coaches in the 50% winning percentage zone than is predicted by math.

We realize that this isn't a truly scientific study – and that some other effects (such as a truly superior team, a great player/leader doing "the work," or other data correlations and skews) may be at work – but the results show that there is indeed "great leadership skill" in the great coaches in sports history.

This is also the reason why you see so many familiar names managing and coaching championship teams. John Wooden and Red Auerbach in years past – and today's great coaches, Phil Jackson, Joe Torre, Bill Parcells, and Bill Belichick – are examples of coaches who are outstanding team leaders.

Quant Fact: All-Time Great Coaches

As we collected data to look at coaches and the effect they might have on team performance in title games, we thought it would be interesting to rank some of the all-time great coaches. Note that the list is a sampling of coaches from the major sports and is not complete or exhaustive. All of these coaches are great, having led their teams to multiple title shots. We are pleased that our measure of skill in coaches seems to agree with the general consensus of the all-time greats.

Please let us know if you feel a coach, whose team has made multiple finals appearances, is worthy of being added to our list. The table shows a coach's win-loss performance in championships, winning percentage, and games behind (GB) the leader. *Note that for college football, the table includes Bowl games that had national title implications.*

On the next page is a table ranking the all-time great coaches.

Table: All-Time Great Sports Coaches
Performance in Championship Games and Series

	Coach / Manager	Sport	Team(s)	W	L	Win %	GB
1	John Wooden	NCAAB	Ind St, UCLA	10	1	0.909	-
2	Phil Jackson	NBA	Chicago, LA	10	2	0.833	0.5
T3	Joe Paterno	NCAAF	Penn St	8	1	0.889	1.0
T3	Toe Blake	NHL	Montreal	8	1	0.889	1.0
5	Red Auerbach	NBA	Boston	9	2	0.818	1.0
6	Bobby Bowden	NCAAF	Fla St	5	0	1.000	2.0
7	Joe McCarthy	MLB	ChC, NYY, Bos	7	2	0.778	2.0
8	Scotty Bowman	NHL	SL,Mon,Pit,Det	9	4	0.692	2.0
T9	Chuck Knoll	NFL	Pittsburgh	4	0	1.000	2.5
T9	Gregg Popovich	NBA	San Antonio	4	0	1.000	2.5
11	Casey Stengel	MLB	NY Yankees	7	3	0.700	2.5
T12	Bill Walsh	NFL	SF	3	0	1.000	3.0
T12	Bob Knight	NCAAB	Indiana	3	0	1.000	3.0
T14	Adolph Rupp	NCAAB	Kentucky	4	1	0.800	3.0
T14	Al Arbour	NHL	NY Islanders	4	1	0.800	3.0
T14	Glen Sather	NHL	Edmonton	4	1	0.800	3.0
T17	Bill Belichick	NFL	New England	3	1	0.750	3.5
T17	Joe Gibbs	NFL	Washington	3	1	0.750	3.5
19	Joe Torre	MLB	NY Yankees	4	2	0.667	3.5
20	Bear Bryant	NCAAF	Kentucky, Alab	6	4	0.600	3.5
T21	Bill Parcells	NFL	NYG, NE	2	1	0.667	4.0
T21	Pete Carroll	NCAAF	USC	2	1	0.667	4.0
23	Sparky Anderson	MLB	Cinc, Det	3	2	0.600	4.0
24	Walter Alston	MLB	Brooklyn, LA	4	3	0.571	4.0
T25	Connie Mack	MLB	Philadelphia	5	4	0.556	4.0
T25	Pat Riley	NBA	LA, Miami	5	4	0.556	4.0
T27	Roy Williams	NCAAB	Kansas, NC	2	2	0.500	4.5
T27	Frank Chance	MLB	Chicago Cubs	2	2	0.500	4.5
T27	Tommy Lasorda	MLB	LAD	2	2	0.500	4.5
30	Miller Huggins	MLB	NY Yankees	3	3	0.500	4.5
31	Mike Krzyzewski	NCAAB	Duke	3	4	0.429	5.0
T32	Dean Smith	NCAAB	NC	2	3	0.400	5.0
T32	Tom Landry	NFL	Dallas	2	3	0.400	5.0
T32	Tony LaRussa	MLB	Oak, SL	2	3	0.400	5.0
T35	Jim Boeheim	NCAAB	Syracuse	1	2	0.333	5.0
T35	Mike Holmgren	NFL	GB, Seattle	1	2	0.333	5.0
37	Don Shula	NFL	Miami	2	4	0.333	5.5
38	Earl Weaver	MLB	Baltimore	1	3	0.250	5.5
39	Bobby Cox	MLB	Atlanta	1	4	0.200	6.0
T40	Bud Grant	NFL	Minnesota	0	4	0.000	6.5
T40	Marv Levy	NFL	Buffalo	0	4	0.000	6.5
T40	Charlie Grimm	MLB	Chicago Cubs	0	4	0.000	6.5
43	John McGraw	MLB	NY Giants	3	8	0.273	7.0

Easy To Coach

Great coaches can not succeed without the right players. And some athletes are very easy to coach. These players tend to be hard working, students of their game. They are the kind of players who remain curious as to how they can improve, grow and learn more about themselves and their sport. They take responsibility for their actions and they relate well to teammates, coaches, the media and people in their personal life.

They also tend to have a balanced life which reflects other interests in addition to the sport they compete in. Having several interests can sometimes help to prevent burn out in young athletes. Also, a well-rounded person may be more likely to handle disappointment in sport, because "all of their emotional eggs" are not in just one basket.

My coauthor and I plan to develop a coachability index in the future. We believe this will be very helpful in judging, recruiting, training and signing the right kinds of athletes for teams in a wide range of sports.

This index may also help to predict which team will perform best in big games. It seems plausible that a team of highly coachable athletes will play better under pressure than a team of athletes who are tough to coach.

Willingness To Assume
And Accept Responsibility

Competitors, coaches and teams who excel tend to be the kind of people who take responsibility for their actions. Doing this allows them to grow, to mature and to learn how to rise to the big occasion.

Conversely, athletes and coaches who blame others and fail to take responsibility for both their successes and for their failures will probably fail to grow into the kind of mature and competent players who are likely to perform well in the big contest.

Respecting Your Adversary

Trash talk and criticism of opponents are sometimes part of big games. And some coaches and athletes use these verbal assaults to psyche out and unsettle their opponents.

However, athletes who "take the high road" and show respect for the other team and for the other competitors tend to be easier to coach than are those who make a habit of putting down the players on the opposing team.

Player-Coach Relationship

I have observed many athletes and their coaches. Some coaches can be tough on players and some are quite gentle. And some coaches treat different players differently at different times.

However, players will almost invariably perform better if they like their coaches and if they have solid relationships with the people who run the team.

On the other hand, if players dislike the coaching staff and the powers that be, they may perform poorly when the pressure of the big game starts to impact them.

For example, a disenchanted player may lack the motivation due to poor relationships with staff members and management. Other athletes may play poorly as part of an unconscious effort to get back at the coach who they do like or who they are in conflict with.

I have seen this happen many times in my practice. I also believe that there is probably a positive correlation between the player coach relationship and performance in the big game.

Because coaches, athletes and management need to be on the same emotional page prior to the big game, Carlton and I plan to study this issue as we work with teams around the country.

Positive Momentum

Momentum is closely related to factors like confidence, team unity and player-coach relationships. That is, when a team or an athlete string together some victories, their confidence almost always grows significantly.

In addition, team members and coaches can start to feel better about one another. So, winning itself can cure a lot of personnel conflicts which can sometimes exist among team members who are getting ready for a big game.

When a team is a cohesive family, the generate a kind of positive energy. This "energy" helps the team to enter the zone as a unit. (We will talk more about the zone later in this book.)

The energy which accompanies positive momentum allows many athletes to think less trust themselves and their abilities more. Once the players have clear minds, confidence which accompany this kind of energy, they are freed up to play to their potential in the big game.

If a player or a team has five or six impressive victories going into the big game, they clearly have some positive momentum on their side. And if these victories have been powerful and impressive performances, a team can be loaded with significant amounts of confidence, unity and positive energy as they approach the big game. This kind of team can have an edge in the upcoming final game.

Negative Momentum

Conversely, negative momentum or a slump can be make it hard for an athlete or a team to play to their potential. Losing streaks can become like cancers for some competitors. They can sap people of all their self-confidence and create huge obstacles going into an important contest.

A team or athlete that has been playing poorly coming into a big game may lack all the positive elements that the team on a winning streak may possess.

A team on a downturn may have to make some significant shifts in their attitude, strategy and behavior to do well in the big event. It is not always easy to make these kinds of transitions when recent history indicates that the team or player is struggling.

Therefore, recent performances appear to create important indicators about the psychological readiness of a team or of a player entering a big game.

Margin Of Victory Or Loss
In The Last 5 Games

Margin of victory is another good way of measuring a team or player's momentum.

That is, victories by large margins, proves that the athlete or squad is playing well and doing a lot of things right prior to the big game.

Conversely, several losses by large margins going into the important event indicate that the athlete or the team is struggling and losing momentum.

Remember, competitors which can dominate their opponents are likely to feel energized and confident as they approach the big game.

Errors

Mistakes in big games are often psychological mishaps. They can be very costly.

In tennis unforced errors can ruin a player. In baseball, errors can cause a team to lose a game they should win. In football, turnovers, foolish penalties and poor clock management all play a crucial role in the outcome of the contest. In basketball, it is very hard to win the big game if you do not protect the ball well and do not execute plays well. In golf, missing the fairway or failing to sink a three foot putt can be disastrous if these mistakes occur late in a final round.

Errors can also disrupt and disturb competitors' equilibrium and cause huge momentum shifts in an important contest.

Players and teams that have a history of avoiding these kinds of mistakes are likely to perform well when they participate in the big game.

Quant Fact: Errors and Championships

Errors, interceptions, and other misplays often decide the outcome of championship games. This factor is so important that a team's ability to execute "cleanly" can help "predict" the eventual champion. That is, if you study a team's performance in key statistics – such as fielding percentage in baseball, or interceptions in football – you can get an idea of which team has a better chance of playing well, and winning, the big game.

The chart below shows the important roles which errors, error minimization, and defense – can play in determining the outcomes of big games from many of the major sports.

Chart: Playing Good Defense and Minimizing "Errors" as Factor (Championship Winning Percentage in Recent Years)

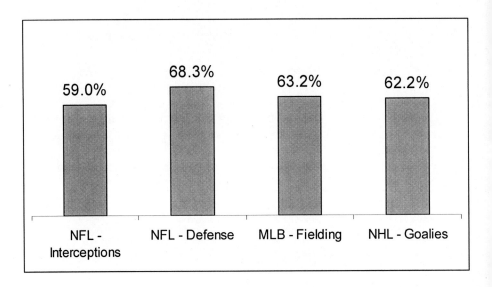

Baseball By The Numbers
Quant Facts: Errors, Consistency, And Winning The Big Game

Baseball, a sport which lends itself to all kinds of statistical analysis, can teach us a great deal about errors, consistency, leadership, and winning. However, before we get into our research on baseball and the role that these factors play in determining the outcomes of big games, we thought it would be useful to comment briefly on one of the enigmas in recent baseball history.

Nobody will argue that the 1991-2005 Atlanta Braves were one of the best teams in baseball history. But why do they have just one World Series Championship to show for their sustained excellence? Is this just a case bad luck? Or can our research on baseball playoffs shed some light on these results? Let's take a look.

Quant Fact: The Curious Case of the Atlanta Braves

During a fourteen-year stretch from 1991 to 2005, the Atlanta Braves finished in first place in their division an unbelievable fourteen straight times (if we exclude the strike-shortened 1994 season). The Atlanta Braves reached five World Series in eight seasons from 1991 to 1999 – but went just 1-4 in these appearances.

The Braves had great teams during this period and won with power and pitching. The Braves' fielding was adequate, but not stellar. **During each of the Braves' five World Series appearances, the team with the better fielding percentage league ranking won the Championship each time.** Luckily, for Braves fans, the 1995 Atlanta Brave team had the second-best fielding in the National League (which bettered the Cleveland Indian fielding) – the year the Braves claimed their sole title during this period.

In addition, we note that the Braves often had good, balanced pitching – but lacked truly great pitchers who were peaking at the right time. We measured this leadership potential by

37

studying the wins for the World Series participants' top two starters. In a Championship Series, it often takes great performances to lead their team and bring home the trophy. *This leadership and competitive spirit is what we try to capture with quantitative analysis.*

Finally, the Braves were also built on power – but our research shows that the "consistency" of "batting average" is very important during the playoffs in baseball. Perhaps during the regular season, a majority of homeruns are hit versus average pitchers (and they are hit during less meaningful regular-season games). These issues might be interesting to study sometime. The old adage that "good pitching beats good hitting" seems true – especially when great pitchers are pitching against power hitters.

There are several key psychological and mental factors impacting "peak performance" and playing "in-the-zone." These factors are especially important at crucial moments like the World Series – and they did not favor the Atlanta Braves.

Quant Fact: In-The-Zone Metrics Predict 15 of 16 World Series Winners

To get an idea of what is going on during World Series play, we went back and studied the results over the past twenty World Series. Baseball has proven to be a good sport to measure some of the key psychological indicators such as peak performance and mental toughness. We have good indicators for many of the key areas where we measure psychological, mental and peak performance. What would you say if we said that these measures of "in-the-zone" performance predicted 15 of 16 World Series winners?

Quant Fact: Mental Toughness and Playing Error-Free

As we did for other sports, we tried to develop a model – or a measure – to quantify the mental frame of mind of the teams. For baseball, we used "errors" to assess how well a team is executing. On one hand, we didn't expect errors in Major League Baseball to have as large of an impact as interceptions had in football.

38

On the other hand, we did not want to overlook the potential damage to morale – as well as the extra out (and runner on base) that the opposition is given. The impact of fielding and errors should definitely not be overlooked. ***Over the past twenty years, the team with the higher ranking in fielding percentage went 12-7 (63.2%) in World Series.*** Although these results do not factor in the expected odds of each team winning the title, fielding percentage is generally non-correlated – or not related – to expectations.

Quant Fact: Leadership in Baseball

The role of leadership by players and by coaches is discussed elsewhere in this book. When you consider leadership in baseball, you think of starting pitchers. Pitchers initiate the start of every play by throwing the pitch. Starting pitchers set the tone for the game – and especially in short series – a few dominating starters can lead their team to a championship.

Since baseball is a team sport, we compared how dominating the World Series' participants' best two starters were in terms of wins (combined victories during the regular season). Over ***the last twenty years, the team with the better two starters (combined victories) won 68.4% of the World Series (13-6).***

Quant Fact: Baseball, Consistency, and Batting Average

Our research on key winning factors frequently uncovers a particular sport's "dynamic" that is related to consistency. Baseball's consistency "dynamic" is batting average. ***The team with the better ranking in batting average won a whopping 76.5% (13-4) of the World Series over the past 20 years.*** To normalize batting averages, especially with regard to the designated-hitter (DH) in the American League, we used the team rank in their respective leagues.

This shows that consistency is another key to winning championships. During times of stress, you don't want to "need" the homerun to win the game. Consistency is important – and the team with the more established team batting average is favored. This also brings up the old adage, "You are only as strong as your

weakest link." You don't want low batting averages to put an end
to a rally.

Quant Fact: Over-valuation and Glitz

As we quantify and measure performance across various
sports, we notice that some measures are actually "negative"
indicators. That is, some statistics don't factor into winning as
much as expected. These statistics often capture the public's
attention and are already factored into expectations.

Homeruns and power-hitters grab media attention and seem
to be over-valued by the public. However, in a short series, the
consistency of hits seems to be much more important than going
for the big homerun swing. In fact, the team with the worse
ranking in homeruns has gone 14-3 (82.4%) in recent World Series.

Quant Fact: Combining Factors = 15 of 16 Winners

Our statistical and factor analysis uncovered a number of
key factors related to sport psychology and winning the World
Series. Most notable of these factors include: fielding percentage,
leadership (in the form of starting pitchers), consistency (in the
form of batting average), and "big game experience." Combining
these factors has predicted 15 of 16 (93.8%) World Series winners.

This is huge evidence that many of these psychological
factors – that we break down and measure quantitatively – can be
very useful. Although these ideas are often overlooked, they are
very important in determining which team is likely to win the
baseball championship.

Errors And Momentum

There appears to be interesting and crucial relationships between errors and momentum. A key mistake can cause one team to gain momentum and another team to lose momentum. Mental errors can really impact one's performance.

For instance, a fielding error in a late inning in a baseball game can frustrate a team and cause them to lose their focus for the duration of the game.

A turnover in a close football game in the fourth quarter can dramatically move momentum, confidence and optimism from one team to another.

An unforced error at a crucial point can determine the outcome of a tennis match and can deplete a player's will to win.

Missing a simple, three foot putt at a key moment can rattle many top golfers.

Interestingly, errors can sometimes give rise to extended losing streaks when players and teams have difficulty letting go of mistakes.

As you can see, errors can play a huge role in big games. When counseling athletes, we try to teach them what they need to do psychologically and mechanically to avoid errors.

Moreover, we also teach them how to be resilient and how to bounce back after a mistake.

In evaluating competitors prior to a big game, pay careful attention to the number and nature of errors in the five games preceding the championship.

Our research on The Super Bowl, perhaps the biggest of the big games, yielded some fascinating results about the role of errors as well as other factors noted in this book.

Football Figures and Facts
Quant Facts: Turnovers, Leadership, Defense, And Experience

We went through every Super Bowl and sifted through data that would hopefully yield information about factors related to sport psychology. The results are based on forty-three games – every Super Bowl from Super Bowl I in January 1967 to Superbowl XLIII in 2009 – and statistics leading up to the big game.

Quant Fact: Super Bowl "Experience"

"Experience" in championships has proven to be a good indication of potential winning performance across all sports. The Super Bowl is no exception to the rule. For every sport we studied, we used the same "indicator" to give credit for championship experience. If a team or player appeared in a championship over the past three years, we gave them "experience credit." *Teams with more experience than their Super Bowl opponent were 14-8 (63.6%).*

Quant Fact: Defense Wins Championships

This commonly-heard phrase, "Defense wins championships," is generally true. When it comes down to "crunch-time," when things really matter – the data suggests that a good defense can beat a good offense. In baseball, the data shows that good pitching can reduce the impact of the homerun ball. *Similarly, during Super Bowls, the team with the higher-ranked defense (Points Against) has won a huge 68.3% of Super Bowls (28-13).* Statistics show that "defensive rank" is more significant than "offensive rank" in predicting Super bowl winners.

Quant Fact: Overall QB Rating Not Important

Surprisingly, the team with the higher-rated QB **underperforms** in the Super Bowl. This is partly because the key to the Championship game is to play with "controlled aggression." You want to win – but need to avoid mistakes. The team with the seemingly better QB may rely too much on the "big play." Interestingly, the data shows similar results to baseball, where teams that rely too much on the long ball under-perform during the World Series.

Thus, a highly-rated QB actually slightly underperforms in the Super Bowl. Picking the team with the "lower QB rating" yields 23-19 (54.8%). This is not a strong factor, but it tells a lot about what may be relevant in predicting the winner of this game.

Quant Fact: The Mistake-Free QB

On the other hand, if the QB normally commits few errors, that is a big plus. *As the team's leader, the QB – and his mistakes – can have a huge impact on the team's morale.* Thus, picking the team with fewer interceptions during the season is a helpful factor. This factor of *"executing at a high level, without making mistakes," is 23-16 (59.0%) during the Super Bowl.*

Quant Fact: Fumbles are Relatively Random

Interestingly, another factor that pops up is "fumbles" – but not in the way you might think it would show up. This might be due to the fact that some researchers believe that fumbles are more "random" than interceptions. On one hand, most interceptions are clearly mistakes by the quarterback. The quarterback may misread a defense, not see a defender, or try and force a dangerous pass.

On the other hand, although a fumble is also unfortunate for a team, many analysts feel that to a large degree, fumbles are the result of "bad luck" – and not "poor skill." Fumbles appear to be less of "a mistake" and more a random act of a "bad bounce."

Super Bowl data, at least, shows this hypothesis to be true. In fact, teams that have more fumbles during the season are slightly underrated (22-14, 61.1%). Because fumbles may be relatively "random," teams that have a lot of fumbles may work on this in practice and consequently be ready to execute well and avoid these mishaps in the big game.

Quant Fact: Combined Super Bowl Factors

We analyzed various football statistics and their ability to predict the winner of the Super Bowl. We focused on factors related to principles of sports psychology and discovered some important results. Interestingly, specific factors such as the ability to execute with minimum errors, and Super Bowl experience, have proven to be good predictors or success. ***Combining these factors for the Super Bowl would yield a record of 28-8 and a winning percentage of 77.8%.***

Quant Fact: The NFL's Greatest Non-Dynasty: The Buffalo Bills

For the six-year period from 1988-1993, the Buffalo Bills were one of the strongest teams in NFL history. Led by Jim Kelly at QB and Thurman Thomas as running back, the Bills reached four straight Super Bowls from 1990 to 1993, when they amassed a 49-15 regular season record. They are one of the great NFL teams – but have no championships to show for their efforts. We note that during each of the four seasons they reached the Super Bowl – the Buffalo Bills committed ***more interceptions*** during the season than their Super Bowl opponent. Our research shows that this factor makes it much more difficult for the offensive leader of the team to pave the way to the Super Bowl Championship. In addition, the Bills ***ran into teams with superior defenses*** each of those four Super Bowl years, another key factor in winning championships. The Buffalo Bills are one of the best "non-dynasty" teams in history. Sadly, for Bills' fans, people remember winners. There's the old saying, "There are no trophies for second place." This is great lesson for athletes, coaches, sports franchises – and competitors of all kinds: when you have the opportunity to win a championship, try to do everything you can to seize that opportunity. Try to optimize your chances of "winning the big game" by both maximizing performance and minimizing errors.

44

Winning And Winning Ugly

While powerful performances can be impressive and can build confidence, there is a lot to be loved and learned from teams and athletes that can find a way to win. These kinds of competitors are resourceful, creative, resilient and determined. They have a special ability to win and often find a way to win games using many of their skills.

Teams that can win in different ways may have more to fall back on in the big game than do teams that have been winning by large margins over and over again prior to the big game.

While winning ugly may be hard to quantify, it is something that is worth paying attention to – when looking at the participants in a big game.

Quality Of Opponents

In evaluating competitors, it is also important to consider the nature and strength of the opposition in recent games. Wins against tough teams or against highly ranked players indicate that a competitor is physically and mentally tough.

Victories against weaker opponents are still wins, but they may not be as meaningful as predictors of performance in the big game.

Quant Fact: Quality Wins as Predictors

Performance and ratings ("power ratings") based on the quality of one's opponents definitely tell a more complete picture of a team or athlete's true "strength" than mere wins and losses. When computing "power ratings," a team's "strength of schedule" and "quality wins" – are important ingredients.

In most sports, teams do not play a "balanced schedule." All teams play a different set of opponents. This means that one team's win-loss record – and overall performance – is difficult to compare with another team's record and tell anything really meaningful.

This is particularly true in sports with fewer games, such as football and college basketball. For instance, it is difficult to compare: (1) a team that has run up a 10-0 record versus a relatively weak schedule with (2) a team that plays a very difficult schedule and has a record of 8-2.

By studying a team's "strength of schedule" and "wins against quality opponents" – you can get a better idea of a team's true strength. Indeed, our work with "quality wins" and other key sport psychology factors presented in this book – have proven to be good predictors in determining playoff winners in the NFL and other sports over recent years.

Strategy

Winning the big game is not just about physical skills. A team or person that prepares better and has a better game plan often wins the big contest.

In addition, a good strategist has the ability to surprise and confuse his adversary by doing something different than what they may have been anticipating.

Similarly, a good tactician has the ability to launch an effective counterattack when they are faced with something new or something unexpected.

In a team sport, the development of a game plan or strategy is usually the responsibility of the coaching staff or the manager. In an individual sport the coach and the athlete often collaborate in order to develop the right strategy. As you can see, strategy is closely related to the coaching and leadership factors described at the start of this book.

When evaluating two competitors, you must consider which coach or which athlete is more intelligent, more creative and more effective where strategy is concerned.

It is not always the best athlete that wins the big game. Sometimes, it is the smartest competitor who takes home the trophy.

Genetics
And
Mother Nature

While this book is primarily about the psychology of sports, you simply can not ignore the fact that genetics and mother nature play important roles in the development of athletes.

Some kids are born stronger, faster, more agile and with better hand eye coordination than others. This is not to say that an athlete can not improve with training and hard work. However, height and size are god given assets which are usually present in one's genes.

If you want to be a lineman in professional football in today's, you probably need to be over six feet tall and weigh over two hundred and sixty pounds.

Similarly, if you want to be a jockey, you probably need to weigh under one hundred and fifteen pounds.

Many of the top athletes I have coached come from athletic families. In fact, I have counseled lots of sons and daughters whose parents were outstanding athletes. Athleticism seems to run in families. Interestingly, recruiters at the collegiate level like to attract athletes who have "athletic genes" in their family.

I believe that genetics and modern science will teach us a great deal about the roles that heredity and mother nature play in the development of athletes who can win when the pressure is on.

Quant Fact: Genetics, Mother Nature, and Champions

Do genes impact one's ability to win championships? Perhaps. This question would make for a very interesting research project. For now, however, we point out the fact that some of sports' brightest stars are indeed blood-related:

- Quarterback brothers, Eli and Peyton Manning
- Baseball's DiMaggio brothers

- Baseball's father and son duo, Ken Griffey and Ken Griffey Jr.
- Baseball's father and son duo, Bobby Bonds and Barry Bonds. Barry Bonds was one of the baseball's greats – even before the steroids.

Speed, Quickness,
Reaction Time Anticipation

Speed, quickness and reaction time play a very important role. In track, football, tennis and basketball, for example, the faster athlete or team has a huge advantage. And there is often little that the slower athlete can do in these sports to neutralize the faster opponent.

Therefore, in analyzing and comparing two opponents, you must take note of the faster performers going into the big game.

Interestingly, some teams and athletes appear to be quicker because hey anticipate what their opponent is going to do rapidly. This skill allows boxers to see punches coming early. It allows baseball hitters to guess right on the kind of pitch. Anticipation allows tennis players to move toward the ball early. Good anticipation allows defensive players to know what the offensive player is about to do.

So, anticipation is tied in with speed, quickness and reaction time.

These are important skills which can impact the outcome of many big games.

Knowing How To Relax

You can not accomplish very much in sports or in life if you are overly tense or if you are exceedingly anxious.

Great competitors know how to quiet their minds and their bodies before the event begins and during the event.

An athlete who is too up tight may not be able to calm himself or herself down enough to perform during the big moment. Others calm themselves as the game moves along.

Most fine athletes have taken training in self-hypnosis, visualization, guided imagery or meditation. In fact, almost every athlete who comes to see me wants and benefits from this training at some point in the process.

While athletes don't want to be lethargic, it is desirable for them to be calm when they compete because calmness is an integral part of being in the zone.

Formal training in relaxation procedures seems to help many athletes to manage the stress of the big game and perform their best. Twenty three of these techniques are included on How To Get Into The Zone With Sport Psychology And Self-Hypnosis. This two CD program is available at www.StayInTheZone.com. Or you can call 888 580-ZONE.

Relaxation, Anxiety And Confidence

These three factors have a dynamic relationship. Based on my work with many athletes, it appears that when anxiety goes down confidence and relaxation rise.

Similarly, when anxiety rises, confidence tends to go down.

When confidence goes up, relaxation tends to elevate and anxiety tends to be lowered or minimized.

Levels of Relaxation and Anxiety Impact Confidence Levels

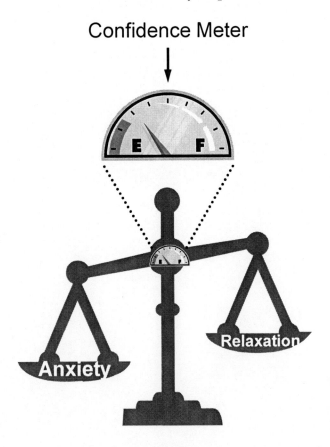

It is sometimes hard to tell which occurs first in a given competitor, since these three factors are so closely connected to one another.

Confidence, Relaxation & Anxiety Have a Dynamic Relationship

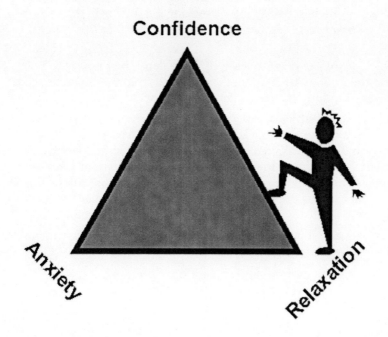

Relaxed Aggression

Interestingly, many sports require the athlete or the team to be able to create a state of relaxed aggression. Think about the golf swing, or the baseball swing or striking a tennis ball. Even boxers and martial artists can benefit from performing in this state of mind.

Athletes need to learn how to blend their ability to relax with their ability to channel their energy, strength, quickness and power. Finding this particular mental gear is not always easy.

Some athletes, however, do this quite naturally. Others can learn how to do it through mental training. Many of my CD programs and books for athletes have sections intended to help athletes find the right level of relaxation and the right level of aggressiveness. I sometimes encourage athletes to think of this as a particular mental gear which they can discover and activate for the big game.

Look for athletes who compete with relaxed aggression, or "controlled aggression," in big games.

Focus

Great teams and great athletes have the ability to concentrate intensively in competition and in practice. They have learned how to eliminate distractions and zero in on what is essential at the moment.

An athlete with an overactice mind will have trouble playing well in a major event. This clutter will harm his or her performance.

Consequently, I frequently teach many athletes to approach their sport with just one idea in their mind at a time. For instance, a baseball hitter may want to tell himself to hit the ball in the gap. This one idea grounds the athlete and helps him to remove clutter from his mind.

Other athletes are trained to empty their minds and play their sport with what I call the empty mind. This is a bit more advanced, but many athletes love having the feeling of the empty and quiet mind when they compete in an important contest.

Minimizing distractions and attending to what is essential are important skills to assess when evaluating competitors. Athletes with a clear, sharp and precise focus are apt to perform better than competitors who have too much going on their brain at game time.

Golf, Focus And Winning Tournaments

Golf is a sport which is quite mental. It is not as much about speed, strength and physical size as is the case in some other sports. In fact, many petite women can outperform bigger and stronger males because of great technique, balance, tempo and weight transfer.

Once one masters the mechanics of the game, golf becomes quite mental. And to play great golf, you have to be focused, relaxed, confident and optimistic every time you step up to hit the ball. Moreover, to win big tournaments, you have to bring your mental game to the final round of the event.

In addition, you have to have the kind of mental and physical game which helps to minimize costly errors. The mental focus, quiet mind and avoidance of errors which are required for success in golf play important roles in other sports too. Tennis, bowling, shooting, archery, pitching, billiards, gymnastics, diving and ice skating are good examples of sports which require this kind of mental clarity.

Carlton's data on professional golfers shows the crucial role that these elements play in tournament golf. (We plan to study some other sports in the near future.)

Deconstructing Golf, By the Numbers
Quant Fact: What Makes a Champion Golfer?

If we merely used "Scoring Average" – we would have a great predictor of future success – but we would have little information on how to win a golf tournament. In this "knowledge discovery" project, we are trying to gain insight into the factors that can improve our chances of winning. Therefore, we attempt to break the overall game down into components that will be helpful in improving performance.

We analyzed 2009 data for the PGA as well as historical data for two of the best golfers in the world today: Tiger Woods and Phil Mickelson. The table below shows the top PGA Money

Leaders in 2009 as well as some of their key statistics. We included data on professional golfers, staggered down to the #150th ranked golfer, in order to study the factors related to success in golf.

Note that we highlighted three columns in the table. These columns show the factors that are most related to success in professional golf based on a statistical measure called correlation:

- Birdies Average (or Bogeys)
- Putts
- Greens in Regulation.

At the professional level, these factors are not very related to success in golf:

- Driving distance
- Driving accuracy
- Eagles

The mathematical results are both fascinating and gratifying because they show that a large portion of tournament performance can be coached and taught. God-given talent and skills such as "driving the ball far" are much less important than putting – and avoiding errors, bogeys, and bad rounds. These parts of the game, (putting and reducing errors), are quite mental and can more readily be studied, coached, practiced and improved upon when the right interventions are made.

Table: 2009 PGA Money Leaders and Rankings
in Various Categories

2009 Rank	Player	Tourn	Money Earned	Drive Distance Rank	Drive Accuracy Rank	GIR Rank (3)	PUTTS GIR Rank (2)	Birdies Avg Rank (1)	Eagles Avg Rank
1	Tiger Woods	17	$ 10,508,163	21	86	16	23	1	7
2	Steve Stricker	22	$ 6,332,636	104	53	57	1	7	101
3	Phil Mickelson	18	$ 5,332,755	13	179	127	35	28	8
4	Zach Johnson	26	$ 4,714,813	143	10	28	34	36	11
5	Kenny Perry	24	$ 4,445,562	48	49	36	92	43	118
6	Sean O'Hair	23	$ 4,316,493	46	123	27	113	21	64
7	Jim Furyk	23	$ 3,946,515	152	25	98	21	32	122
8	Geoff Ogilvy	20	$ 3,866,270	63	151	128	29	21	131
9	Lucas Glover	26	$ 3,692,580	25	70	88	51	13	177
10	Y.E. Yang	23	$ 3,489,516	67	133	76	113	53	27
15	Dustin Johnson	25	$ 2,977,901	3	169	80	29	6	10
20	Padraig Harrington	20	$ 2,628,377	106	171	175	104	129	97
25	Ian Poulter	17	$ 2,431,001	133	75	178	61	114	162
50	John Mallinger	27	$ 1,717,140	164	31	115	46	114	44
100	Ted Purdy	30	$ 838,707	45	89	71	89	25	94
150	Ryan Palmer	26	$ 454,510	26	128	104	61	101	52

Note the importance of:

(1) Birdies Average
(2) Putts GIR
(3) Greens in Regulation Rank

"Eyeballing" the 2009 Money Rankings and the three labeled columns (1), (2), and (3) show that these categories are generally more correlated with success. An actual mathematical "correlation" verifies that these are the most important factors. Other factors (such as Eagles Ranking or Driving statistics) are relatively non-correlated to 2009 prize money.

Quant Fact: Birdies and Bogeys, Not Eagles

We studied professional golf "score-related" data such as Pars, Eagles, Birdies, and Bogeys "rates." Interestingly, "eagles" proved to have a very low correlation to ultimate success. In addition, par "rates" – are fairly constant amongst professionals – so that birdies and/or bogeys yielded the most useful information.

These two statistics are correlated fairly well. For pro golfers, birdie information is slightly more useful (in particular, birdie rate or birdies per round) in predicting success. For more mortal golfers, bogey rates (bogeys per round) are crucial. Either way, birdie and bogey data prove to be very similar. *That is, the ability to avoid bogeys – and ultimately bad rounds – is what leads to championships and golf titles.*

It is interesting that even at the professional level the ability to avoid mistakes is such a quantifiable key. This is even more so for golfers who are not professionals.

Quant Fact: Practice Putting, Put the "Driver" in the Back Seat

A statistical analysis of the 2009 PGA Money Leaders data shows some interesting results. Driving has a little-to-no correlation to tournament success, but putting is a real factor. *This is why golfers say, "Drive for show, putt for dough."* You also need a generally good all-around game before you can putt – so that "Greens in Regulation" pops up on our statistical screen.

Interestingly, "Driving Distance" and "Driving Accuracy" are negatively correlated to one another – even at the professional level. That is, on average, golfers who drive the ball far aren't as good with accuracy – and vice versa. Driving in golf (both distance and accuracy, combined) is not very related to tournament golf success. These results are based on a statistical analysis of the 2009 PGA Money Leaders and associated golf data.

Quant Fact: Tiger Woods Over the Years

If you look at Tiger Woods' results over the years, you will see that he has been remarkably consistent. We wanted to highlight a few of his ups and downs in order to show the importance of the major factors we discussed above.

- The year 2000 was one of Tiger's best years. He won 9 of the 20 tournaments he entered and had 17 "Top 10"

finishes. He was hitting on all cylinders, with a #1 ranking
in both Birdies and Greens in Regulation – and a #2
ranking in putting. It doesn't get much better than that.

- 2001 was a slightly off-year for Tiger Woods, with only 9
"Top 10" finishes out of 19 tourneys. This can be blamed
on his putting, as his putting ranking dropped to #102.

- 2004 was another off-year for Woods, with only one
tournament win to his credit. That year, Tiger Woods saw
his Greens in Regulation ranking dip to #47.

Mental toughness and focus are some of Tiger Woods'
greatest strengths. Take a look at Tiger's ranking in "Last Round
Score" over the years. Not only is he near the top, he is regularly
#1 on the list! The stats show that when it counts, Tiger Woods
can definitely "bring it."

Table: Tiger Woods Performance and Rankings
2000-2009

	Tourn.	Top 10	Win	GIR Putts	GIR	Birdies	Eagles	Score Last Rd
2009	17	14	6	23	16	1	7	1
2008	6	6	4	na	na	na	na	na
2007	16	12	7	4	1	1	2	1
2006	15	11	8	35	1	1	2	1
2005	21	13	6	5	6	1	16	1
2004	19	14	1	2	47	2	141	15
2003	18	12	5	10	26	3	1	13
2002	18	13	5	83	1	1	75	1
2001	19	9	5	102	5	5	5	2
2000	20	17	9	2	1	1	1	2
Avg	16.9	12.1	5.6	29.6	11.6	1.8	27.8	4.1
StDev				37.6	15.8	1.4	48.6	5.6

Table: Phil Mickelson Performance and Rankings
2000-2009

	Tourn.	Top 10	Win	GIR Putts	GIR	Birdies	Eagles	Score Last Rd
2009	18	7	3	35	127	28	8	114
2008	21	8	2	22	70	7	79	26
2007	22	7	3	15	84	17	4	23
2006	19	8	2	5	21	2	12	92
2005	21	9	4	15	46	2	144	70
2004	22	13	2	43	10	3	71	23
2003	23	7	0	14	108	13	54	147
2002	26	12	2	5	46	2	4	3
2001	23	13	2	2	26	1	1	55
2000	23	12	4	3	43	4	2	23
Avg	21.8	9.6	2.4	15.9	58.1	7.9	37.9	57.6
StDev				13.9	38.4	8.8	48.1	47.2

Notes:

- Annual performance is correlated to Birdies, GIR Putts, and GIR.
- Good and bad years are explained by performance in these three areas.
- Notice Tiger Woods' consistently good performance during the Last Round.

Quant Fact: Phil Mickelson Over the Years

Phil Mickelson has been one of the greats and is one of the most fun golfers to follow. Golf analysts talk about Mickelson's risk-return characteristics – and in particular, his risk-taking style of play.

- Mickelson has been relatively consistent with the number of tournaments he has won over the years. For Mickelson, some particularly good years for him included 2000, 2001 and 2004 – years where he had Top 10 finishes in more than half the tournaments he entered. Phil has always been a good putter – and in these years, he was one of the best in the game in two out of the three categories: putting, birdies, and/or Greens in Regulation.

- In 2003, Mickelson didn't win a single tournament all year, mostly because his Greens in Regulation dropped to a #108 ranking.

Managing Big Moments
In Big Games

Different sports have different kinds of big moments. In golf, it can be the last putt on the last hole. In baseball, it can be the last out in the last inning. In football, it can be the final two minutes. In basketball, it can be the last shot to win or tie the game. It is interesting to note how athletes and teams perform at critical times in critical games. We found some very interesting data about big moments in big games from the world of tennis.

A New Sport Psychology Measure for Tennis
Quant Facts: Experience, Focus and "Big Point Performance"

Tennis, as in every sport we have studied, shows that "experience" is very important in determining the winner of a championship. We reviewed Grand Slam Finals results for Men's Singles since the Open Era of tennis began in 1968. Over more than 40 years of tennis – across the four major Grand Slams – the player with more recent experience in that tournament's final went on to a 51-26 record, or a 66.2% winning percentage. Experience counts!

Quant Fact: Style of Play, Unforced Errors and Winners

Tennis commentators often mention the number of unforced errors that a player strokes during a match. Many tennis analysts like to study the number of "winners" minus "unforced errors" as an indication of how well a player is doing. Sometimes we hear about the total number of points each player has won.

All of these statistics tell a story of how the match is going. These statistics can also be misleading – or not tell you much more information than the actual match score. For instance, a player's style makes a difference in the number of unforced errors they hit. A very aggressive player will normally hit many more unforced errors – but also hit a correspondingly greater number of winners.

"Winners minus unforced errors" is a great indication of how well each player is doing. Indeed, some of the tennis greats like John McEnroe and Roger Federer have had matches where this measure is – by far – much higher than that of other tennis stars. However, we also wanted to measure how players are doing at "crunch time."

Quant Fact: Focus and "Big Point Performance"

How can we study the "mental toughness" – or the "focus" – of a tennis player? We wanted to come up with a statistic that would quantify "performance under pressure." We wanted to develop a measure that would study how players performed during key moments of a match.

In tennis, "break points" are the biggest points of every match. The server typically has the advantage in tennis – and breaking your opponent's serve is a prerequisite to winning.

This is why we developed a new and innovative "sport psychology" measure for tennis that we call "Big Point Performance." We decided to take a look at how players performed on these key break points during a match. We take the percentage of Break Point Conversions "for" a player and subtract the Break Point Conversion percentage "against" that player to establish the player's performance on "Big Points" during the match.

Tennis "Big Point Performance" =

Break Point Conversion % "for"
Minus
Break Point Conversion % "against"

Using this measure can tell us about each player's current mental state of mind. More importantly, it can tell us if the player is peaking, at a good level of "intensity" and is very focused.

Quant Fact: "Big Point Performance" and the 2009 US Open Final Upset

We wanted to take our theory out for a test-drive and see how well "Big Point Performance" (BPP) measured mental performance and potentially predicted their current level of focus and intensity. For our example, we decided to take a look at the biggest surprise of 2009 in tennis: Juan Martin Del Potro defeating Roger Federer in five grueling sets in the US Open Tennis Final. Here are the match statistics (please see the following page):

2009 US Open Men's Final
Del Potro Upsets Federer
3-6, 7-6, 4-6, 7-6, 6-2

Match Summary	Federer	Del Potro
1st Serve %	91 of 181 = 50 %	111 of 171 = 65
Aces	13	8
Double Faults	11	6
Unforced Errors	62	60
Winning % on 1st Serve	65 of 91 = 71 %	81 of 111 = 73
Winning % on 2nd Serve	50 of 90 = 56 %	33 of 60 = 55 %
Winners (Including Service)	56	57
Receiving Points Won	57 of 171 = 33 %	66 of 181 = 36
Break Point Conversions	5 of 22 = 23 %	5 of 15 = 33 %
Net Approaches	31 of 47 = 66 %	23 of 34 = 68 %
Total Points Won	172	180
Fastest Serve Speed	129 MPH	138 MPH
Average 1st Serve Speed	116 MPH	116 MPH
Average 2nd Serve Speed	95 MPH	91 MPH

Source: USopen.org

In a close match that took over four hours to play, Del Potro pulled ahead in "Total Points Won" just barely, winning 180 points to Federer's 172. In such a close match, Del Potro had a surprisingly high BPP of +10% (33% "Break Point Conversions for" minus 23% "Break Point Conversions against"). Federer's BPP is correspondingly -10%.

Del Potro's "+10% BPP" shows that he played the big points very well in this match. But, more importantly, how did BPP measure up as a predictor leading up to the 2009 US Open Final?

- In his quarter final match versus #16 seed Cilic, Del Potro amazingly broke serve every opportunity he got, going 8 for 8 on break opportunities, or 100%. He was broken 3 times, out of 8 break point chances, or 38% of the time. Del Potro's BPP was a huge +62% in the quarters.
- In the semis, Del Potro faced Nadal, and had a BPP of 38% - 0% = +38%. Nadal, one of the best in tennis today, didn't break Del Potro a single time, with five break point opportunities. Nadal may not have been 100% healthy – but **we can see that Del Potro was playing the big points "huge."**
- In the quarter finals, Federer posted a solid 36% - 0% = +36% BPP over #12 seed Soderling.
- In the semis, however, Federer actually scored a -3% = 30% - 33% BPP in his match over Djokovich. Federer won the match in three straight sets. However, the match was closer than everyone thought – with the score being a tight 7-6, 7-5, 7-5. Although Federer was broken only one time, **Federer had great difficulty winning the big points in breaking Djokovich.**

Below, we summarize each player's Big Points Performance (BPP) leading up to, and including, the Championship Match. As you can see, Del Potro was playing very good tennis on the key points. Federer actually suffered a negative BPP in his semi-final match, suggesting that he wasn't at his mental best. **Maybe Del Potro's big upset was not such a big surprise after all.**

Table: A Look at BPP (Big Point Performance)
Leading Up to the 2009 US Open Men's Finals

	Federer	Del Potro
Quarter Finals	+36%	+62%
Semi Finals	-3%	+38%
Finals	-10%	+10%

Quant Fact: Moral of the story?
Focus and Concentration

We cannot underestimate the importance of focus and concentration at key moments in sports – and indeed, in life. Athletes may not be able to "get psyched up" all of the time – but it is important to practice this "skill" so that you are able to raise your game to the "next level" as the stakes get higher (for instance, championships!) and especially during key moments in a match.

Distraction Control

A concept closely related to the idea of focus which was discussed a bit earlier in this book is distraction control. Athletes have a ton of potential distracters on and off the field.

During a game, they have to deal with crowd noise, the media, trash talk, conflicts with teammates, conflicts with coaches, pressure to perform, desire for playing time, injuries, weather conditions and bad calls.

Off the field, they have to manage their relationships, family life, travel, the press, long seasons, agents, and hours of challenging practice and cross training.

In the big game, the athletes and the coaches must all be able to tune out distractions and tune into their goals and their game plan.

Some competitors are better at this than others. Others need to learn how to eliminate or manage events which can take an athlete out of his or her game. I have taught many athletes how to use their minds to remain focused on what is important.

Controlling distractions in the big game may be something that experienced players do better than do less experienced players.

Observe opposing plays and notice which ones seem to be most business like and most centered during the course of big games.

Carlton and I plan to investigate the roles of both distraction control and focus among athletes and teams in the big game. We suspect that players and coaches who focus better are apt to perform better.

Playing In The Present

Athletes who spend too much dwelling on the past or too much time worrying about the future can not stay in the here and now when the pressure is on.

While it is common for players to berate themselves for a bad play or a mistake, they need to learn how to let go of this error quickly and efficiently in the big game.

Similarly, an athlete who worries about the future can remove himself or herself from what needs to be done now. Some golfers I have coached think about the last hole when they are playing the fifth hole.

A quarterback I worked with would worry about the fourth quarter before he had completed the first half.

Conversely, an athlete who can play in the present will quiet his or her mind and allow their bodies to do what they have been trained to do.

Being in the present is another part of being in the zone and some athletes seem to do this better than others. Players who understand that the most important shot or play is the next one, are likely to play well in the big game.

Visualization, Self-Hypnosis, Guided Imagery, Meditation, And The Zone

Virtually every athlete who I have counseled benefits from learning some form of visualization, self-hypnosis, guided imagery or meditation.

Athletes seem to take to this training very easily and they tend to derive huge benefits from these mental skills. Because these methods help athletes to relax, focus, use their imagination and their dreams, this kind of training is quite helpful for players who are preparing for an important contest.

In my view, the mental states induced by these techniques help athletes to get into the zone and stay in the zone more often. In fact, the zone closely parallels or matches a hypnotic trance or a hypnotic state of mind.

The zone can be thought of as a state of mind where the athlete is calm, confident, focused, is in the here and now and is absorbed in the game or the task. In addition, the player feels that physical actions happen automatically without any thought. The game feels easy. There is no self-criticism and there is an increased belief that your dreams can become realities. Not surprisingly, the athlete is having fun in this mental state.

(I will talk more about the importance of fun shortly.)

Competitors who learn techniques can get themselves into the zone more often. And players who can do this have a big edge over competitors who have not had this kind of training.

I have written several books on this topic and developed programs for athletes in a wide range of sports. Visit StayInTheZone.com to learn more about these materials.

Carlton and I hope to do some studies to determine to which specific mental training methods are most efficient when it comes to improving performance in big games.

Fun

No matter what level an athlete is competing at, it is essential that they love and enjoy their sport and that they know how to have a good time while they compete.

Levity and some humor help to relax the athlete and this fun loving state of mind enables many sports icons to enter the zone more often.

You can be intense and focused and still be able to have a good time in the heat of battle and during the big game.

I have reminded many athletes to realize that are fortunate to be able to earn their livings at a sport they love.

In evaluating competitors, determine which team seems to be having the most fun getting ready for the game and while playing the game.

Resiliency

The ability to bounce back from adversity or from a set back is a wonderful trait for athletes. Some athletes collapse when they hit a valley. Others find another gear and elevate their game. The players who have this very special quality do not panic when they are behind in a big game. In fact, they may elevate their games to meet the challenge of the needed comeback.

Teams or players with a history of comeback wins are often well-suited to win the big game.

Ability To Bounce Back Quickly

Events and momentum swings can happen rapidly in athletics. Athletes who know how to gather themselves and refocus quickly are very valuable to a franchise during a big game.

Conversely, an athlete who gets disappointed and feels defeated after a setback can be a liability to a team. Observe competitors carefully and favor those who bounce back quickly.

In one of my books, *Get Into The Zone In Just One Minute*, I present a number of exercises for regrouping and refocusing in a very efficient manner. This book is available at www.amazon.com and at www.barnesandnoble.com.

Managing Emotions

Some athletes celebrate too much when something good happens. Others get pretty down after experiencing a disappointment. Athletes who can avoid these emotional peaks and valleys are apt to play more evenly and more consistently.

I encourage my clients to avoid getting too high or too low prior to the big game.

Furthermore, a player or team with the right level of arousal or intensity is apt to be the victor in the big game.

Anger And Frustration

Frustration and anger are closely related emotions. Some players report that frustration precedes anger.

Players and coaches need to be able to handle these feelings well if they are to perform well in the big game. Even boxers and martial artists have to control these kinds of emotions.

I have taught many athletes mental techniques for managing this feelings. Some of them benefit from learning how to release anger and frustration. Others learn to moderate and to channel the energy connected with their anger in a positive manner during the game.

Competitors who can manage these powerful feelings can have an edge when they experience disappointments, conflicts or setbacks in a big game.

Consistency

As was just noted, a player who can manage his or her emotions is likely to perform in a predictable and consistent manner. These players approach their craft in a predictable manner. The know what they can do and they know what they can not do.

Consistent teams have a successful game plan that they adhere to over and over again.

Consistent, reliable players also seem to have a positive impact on their teams and on the players around them. Derek Jeter, Oscar Robertson, Joe Montana and Peyton Manning are excellent examples of consistent performers who help their teammates to play in the zone during a big game.

Consistent players who are team leaders can be huge assets in the big game. A team with two or three players like this can be quite dominant in an important game or series.

Quant Fact: Consistency Wins Championships

The importance of "consistency" in winning championship titles should not be underestimated. Sports fans often remember a star like Reggie Jackson who carried the Yankees to the 1977 World Series title by hitting five homeruns and batting .450. In baseball's World Series, however, "consistency" in the form of "hits" is more important than the threat of the home run. Just think: we had to go back to 1977 for one of the most memorable and dominating performances by a slugger.

In baseball, we learned that the team with the better ranking in batting average won a whopping 76.5% (13-4) of the World Series over the past 20 years. Similarly, in the Super Bowl, we discovered that one of the main differentiating factors is a quarterback who performs at a consistent, error-free level – and not necessarily the flashy high-rated quarterback.

More often than not, our research has shown that "consistency" is one of the major factors in determining the winner of the big game in any sport. Consistency, whether it be a continued rally in baseball – or solid, error-free performance by a football quarterback or hockey goaltender – improves a team's morale and winning results, in a measurable manner. Team sports are more of a "team" sport than you think!

Hard Work

I have coached some very talented athletes who felt that they could excel in their sport because of their outstanding natural ability. Most of these athletes get a wake up call at some point in their career.

The successful ones recognize that it takes a great deal of hard work, learning and practice to get to the top of any game and to be ready for the big game when it comes along.

A player or team who works hard and practices diligently is likely to be ready for the big challenge when it comes along.

Quant Fact: Practice the Less-Exciting Skills to Win

Offense is the more exciting side of sports. People enjoy the skill involved in throwing a TD, hitting the long ball, or hitting three-point shots under pressure. There is a lot of inborn talent that leads to offensive skills – that helps to win sporting events.

Equally as important, however, are the less-glitzy skills. Our data shows that factors – such as team defense, improving fielding percentage, and minimizing errors or turnovers – although less exciting, are very important in winning big games.

These skills, which are connected to hard work and practice, can be the deciding factors when winning games, sporting events, or championships – at any level. We have seen that these more "everyday" skills, which include fielding or batting average in baseball or putting in golf – are crucial to winning championships.

Interestingly, these are the kinds of factors that can often be improved and honed through the kind of diligence we see amongst great players and great teams prior to competing in big games. Below are a few more thoughts on the psychology of practicing.

Practice

In most instances, players and teams that practice well, tend to play well. There are situations where players look bad in practice, but play well during the big game, but this, in my experience, is rare.

There are a number of elements which comprise good practice. Hard work is one of them. Simulating game situations is another. Devoting a lot of time and energy to what the player or team is weakest at is vital. Practice prior to the big game also needs to connect to the strategy which is planned for the big game.

You can tell a lot about an athlete or a team's readiness by watching how they practice prior to the big game. You can get a sense of their energy, tension, camaraderie, confidence, relationship and focus.

If a team is not practicing well, something needs to be fixed and changed prior to the big game. Coaches, managers and sports pundits can learn a lot by watching teams practice prior to the championship game. The competitors should be peaking as the big game draws closer.

The Intelligent Way To Practice

Athletes who can win the big game tend to approach practice as an opportunity to prepare, improve and to master what they need to know for the big game.

I have observed lots of athletes in practice and I am partial to those who remain curious about their game and themselves and who continually ask themselves what they just learned from the last swing, throw, kick, play or maneuver.

Players who approach practice with this openness empower themselves to continue to grow and improve at their craft.

Conversely, players who practice poorly with too much bravado or too little enthusiasm don't give themselves and their team a chance to win in a big contest.

As I mentioned earlier, players who practice well tend to play well when they are faced with a big challenge.

Playing At Home, Playing Away From Home

Lots of athletes play better at home. The familiar surroundings, the comfortable routine and the energetic and supportive fans clearly impact the way players feel during the big game. Given the choice, most players and coaches would rather play at home, so you must consider where the team is playing when comparing one team against another.

If the teams are playing on a neutral field, the number and energy level of their fans in the stadium or the arena can have an impact on the competitors and on the outcome of the game.

Quant Fact: Home and Away

Home field advantage is consistent across every sport. It is well-documented that professional and college sports teams perform better at home. Interestingly, this home/away impact even shows up on individual statistics. We recently performed some historical research for MLB players – and noticed this bias to be consistently true over the years. In recent years, players have batting averages that average about 10 points higher at home than on the road.

One big example of a player who enjoyed the benefits of playing at home was Wade Boggs, who hit almost 50 points higher at home during his career! And – the impact wasn't true only at Fenway Park, with its Green Monster. When he played with the Yankees, Boggs hit about 40 points higher at home. Wade Boggs, known for his diet of chicken – certainly liked that home-cooked chicken!

Goals And Goal Setting

Athletes, coaches and teams spend a lot of time talking about goals. And clear performance goals and psychological goals help competitors to stay focused and motivated as they approach the big game.

Well focused athletes and cohesive teams tend to describe the same goals and the same objectives in the same way. When you get a sense of this kind of single mindedness from the personnel, you know that the athletes are all working together for the same thing in the same way.

Everyone has bought into a way of playing, a way of preparing and a way of communicating with the media. Teams with common goals tend to perform well during the course of a season and in the final game.

Conversely, teams with selfish players with different sets of goals tends to have trouble winning consistently in the games leading up to the big game and in the big game.

Coaches often have a lot to do with the goal setting in the same way that they have a lot to do with strategy development and team building.

When competitors know what they are trying to accomplish and what their teammates are working towards and when they have a good sense as to what they need to reach these goals, they are apt to be well prepared for the big game.

If they are focused and enthusiastic, these are also good indicators as to how they feel about their goals.

The Locker Room

Some locker rooms have a family atmosphere where you can sense great camaraderie, great energy, a sense of purpose and outstanding team chemistry. Coaches, captains and the personnel have a lot to do with the creation of a winning atmosphere in the locker room and during the big contest.

Other locker rooms are cool, distant and lifeless. You can sense that the athletes are distant from one another, in conflict and somewhat selfish.

The team that has a family feeling but is also disciplined is apt to have the closeness and sense of purpose which is required to play well when they need to manage the pressure of the big game.

I have been in a number of different locker rooms and it is usually pretty easy to get a sense of the team chemistry by spending a little time around the players. The locker room atmosphere tells you a lot about how ready a team is for the big game.

Similarly, you can learn a great deal about the mental toughness of an individual athlete by observing them in their locker room prior to a big match. Boxers, fencers, martial artists and golfers usually have a routine that they adhere to as they get ready for their big game.

Whatever they do, it is essential that they feel physically and psychologically comfortable and ready to go when they leave the locker and get ready for their big challenge.

Stars Who Can Lead

Outstanding players can obviously help a team to win big games. However, it appears that great players who can act as team leaders can have a huge impact on a team playing in a big game. Magic Johnson is an excellent example of a great player who had a positive impact on his fellow players on the court and in the locker room.

We found some interesting numbers about leadership, and big games from the sport of professional hockey.

What Do The NHL Numbers Say?
Quant Facts: Star Leadership and "Finals" Experience

The results show that leadership is crucial during the Stanley Cup Finals. In the NHL, we mean leadership at both ends of the ice. For the purposes of our NHL research, we took a look at the past 30 Stanley Cup Finals, from 1979-2009 (note that 2005 was the year of the NHL strike). First, let's take a look at defensive leadership: the goalies.

Quant Fact: Goalies As Leaders

"The puck stops here" is a common slogan. If the other team can't score, you can't lose, right? The team that wins the Stanley Cup is often the team with the "hot" or "more focused" goaltender. The goalie has the potential to step up as a leader and to be the backbone of the team. The "in-the-zone" goalie will see the puck like it is a basketball and lead his team to a championship.

During "typical" rough-and-tumble playoff hockey, defense and a strong goalie can lead a team to the Stanley Cup Championship. The data *does* show that defense wins championships in the NHL.

If you compare each team's goaltender based on save percentage, you can see which goalie has the better potential to lead their team to The Stanley Cup. Note that *"save percentage"* yields better information than "goals-against-average" because playoff hockey is a little different than regular season hockey. The game tightens up – and the tempo changes somewhat – so that a pure save percentage works slightly better. ***Save percentage has been 18-11 or 62.2% over the past 30 years.***

Quant Fact: NHL Offensive Leadership

Similarly, if you rate each of the team's best offensive players by points, you can see which team has the potential for their offensive leader to make a difference and take control of the game. ***Actions speak louder than words and good leaders lead by action.***

Leaders on the offensive side performed particularly well during the high-scoring days of Wayne Gretzky. Besides that high-scoring era, goalies that are "in-the-zone" seem to be able to lead their teams to championships more easily. ***Offensive leaders as a standalone indicator over the past 30 Stanley Cup Finals yielded a record of 19-11 or 63.3% winning percentage.***

Quant Fact: NHL Finals "Experience" and Overall Results

As we have seen in other sports, experience plays a huge role in Stanley Cup Finals. We gave a team Stanley Cup Finals "experience" if they played in a Final over the previous three seasons. Using this indicator as a stand-alone system would yield 11-2 or 84.6%. Repeat: 84.6%; that is a huge number!

Combining these indicators results in a 15-1 (93.8%) record in predicting the winner of the Stanley Cup Finals. ***Yet again, this is strong evidence that leadership and experience play very strong roles in determining who will win the big game.***

Quant Fact: Stanley Cup Finals and the Rise of the Offensive Leader

Although hockey is very much a team sport, leadership at both ends of the ice is important. Generally, during hard-checking and physical playoff hockey, defense and goalies who are "in-the-zone" are major determinants of the eventual victors.

The exception to the rule is when you have standout offensive leaders like Wayne Gretzky. "The Great One" ushered in a period of high-powered NHL scoring from the mid-80's to the mid-90's. During this time, offensive leaders were more easily able to "lead" and "will" their way to championships. Gretzky, led his Edmonton Oilers to several Stanley Cups, and then Mario Lemieux did the same for his Pittsburgh Penguins.

The chart below shows the importance of "stars who can lead" in the NHL as well as some other major sports. The team with the potentially stronger leader(s) – has a materially better chance of winning the big game.

Chart: Star Player "Leadership" as Factor
(Championship Winning Percentage in
Recent Years for Team with Stronger Leader)

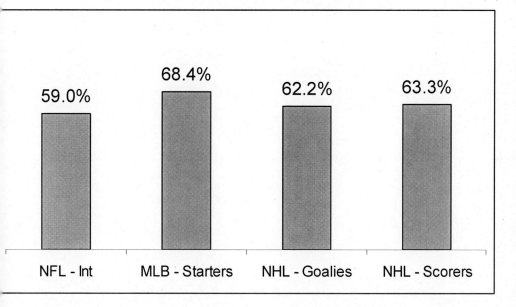

How Many Super Stars Do You Really Need?

I believe you need at least two super stars at key positions to win championships. And having three outstanding players is probably better if you want to be the best team in most sports.

Look at great teams and think about it. They almost all had at least two phenomenal players. Here are just a few that come to mind:

- Yogi Berra and Mickey Mantle

- Kareem Abdul-Jabbar and Magic Johnson

- Michael Jordan and Scotti Pippen

- Joe Montana and Jerry Rice

The number of stars who can efficiently help their teams to win in the big game is something we plan to study in the near future.

A Mix Of Older Players And Younger Players

A team that has a blend of experienced veterans and energetic, enthusiastic younger players can be well suited for playing well when the whistle blows to start the big game.

This mixture of players can give rise to healthy competition amongst teammates and it can also set the stage for some positive and useful mentoring where the senior players can teach and encourage the players with less experience.

In looking at teams that are about to compete in a big contest, be aware of the age mix of the team.

Parents, Athletes And Coaches

Where young athletes are concerned, it is imperative that the kids, the parents and the coaches are all on the same page or at least on a compatible page if the player is to perform well in the big event.

I have counseled lots of athletes, coaches and their families in order to strengthen and improve the quality of the relationships within this triad.

If there is conflict among the members of this group, it can have a disastrous effect on the young competitor who is trying to get ready for a challenging match.

Conversely, if all three people are in harmony and are clear on their approach to solving problems, the athlete has a better chance of playing to his or her potential in the big contest.

Sometimes, an athlete's entourage can include an agent, a fitness trainer, a physician, a message therapist, a nutritionist, a mental toughness coach, a manager, a spouse or a significant other. Again, the same principle applies. Everyone needs to be working together as a unified team to achieve the shared goals. This kind of cohesiveness is essential prior to the big game.

Parenting

In sports and in life, it is very useful if you have parents who are loving and supportive and who know how to set appropriate rules and limitations with their children. A solid family background can help to develop a fine athlete who is also a solid citizen.

Now, many fine people do not grow up in this kind of loving and well-structured environment and they turn out fine.

However, this type of parenting can increase the likelihood that a person will grow and mature in a positive manner.

Conversely, bad parenting can hinder a person's development and block that person from reaching his or her full potential is sports and in life.

Now, many fine athletes come from single parent households, so this need not be a problematic issue. The main things is that the person feel loved, supported and well supervised while he or she is growing up.

I have seen the impact of parenting in my practice thousands of times. Many parents make the mistake of not setting enough limits with youngsters. And some talented athletes get special treatment because of their athletic gifts.

Not surprisingly, an athlete who was spoiled and overindulged can be very hard to discipline and hard to coach.

Also, an athlete who was criticized a great deal can lack the confidence and self-esteem they need to perform well consistently in important contests.

So, when evaluating an athlete, it is crucial to get a sense of the family environment they were raised in. A dysfunctional atmosphere while growing up can produce an athlete who is tough to coach and tough to get mentally ready for a big game.

And as was suggested earlier, solid parents can help to create a solid athlete and a stable person. The connections between parenting and the development of athletes is something that Carlton and I plan to investigate in detail shortly.

Parent-Coaches

Some of the top athletes I have counseled were introduced to a sport by one or by both of their parents. In some instances, their mother or their father was their first coach. This kind of connection with a parent can provide an athlete with a lot of good feelings about the sport and it can serve as positive foundation for the psychological development of the competitor.

Conversely, if young athletes have a conflicted relationship with a parent around their sport, their development can be significantly hampered.

So, in evaluating one's ability to perform well in big games, it is useful to explore and to understand their relationship with their parent-coaches.

Relationships With Siblings

Brothers and sisters often introduce their younger siblings to sports. And siblings can play an instrumental role in developing a young athlete. Many of my patients recall fond memories of their early experiences competing against or with a brother or a sister.

Siblings, like parents, can be powerful mentors for the emerging athlete. Being mentored by an older brother or sister can help a child to master the physical and mental aspects of a sport.

I would encourage coaches and managers to be familiar with the kinds of relationships players had with their siblings when they were kids.

Believe it or not, these early relationships can have an effect on big game performance.

Body Language

Confidence, composure, focus, sense of purpose and relaxation are all communicated through body language. Athletes who stand tall and seem centered and balanced are probably in a good frame of mind to perform well.

Conversely, athletes with slumped shoulders who fail to make eye contact and are shake their heads are telling you that they are frustrated, without focus, lacking confidence, off their game and "out of the zone."

Observe athletes prior to game or during an interview and see what their body language is communicating to you, to their teammates to their coaches, to their opponents and to others.

Favor athletes, coaches and teams who look like champions on and off the field and who appear to be ready to go and "in the zone" as the big game approaches.

Demeanor And Routines

Watch Derek Jeter at the plate. He always has the same routine and ritual. If you observe him very closely, you will see that you can not tell whether it is his first at bat or his last at bat in the bottom of the ninth inning to win a big game. He looks the same.

Bjorn Borg, the tennis great, was the same way. He looked and acted the same way between every point. You could not tell what was happening in the match by watching him. He stayed cool and calm no matter what was transpiring around him and in the match.

Peyton Manning, the outstanding quarterback, is very similar. He approaches every snap from center with the same attitude and demeanor.

Walt Frazier the basketball legend was always smooth and even-tempered. He also looked the same no matter what was on the scoreboard.

And Fred Couples, the outstanding golfer, seems to approach every shot and swing the same manner.

Great athletes keep these routines to stay calm, confident and focused. These rituals help to keep them in the zone. In addition, they mask any self-doubt or negative thoughts from the competition by keeping their uniform demeanor and body language.

Look for players who approach the big game in this stable and predictable manner with their consistent routines and rituals.

Managing The Media

The media attention surrounding a big game can be a distraction for some athlete and for some teams.

Others enjoy the attention and handle the contact with the press with grace. I do not believe that you must be a media star in order to perform well in the big game. And some athletes get better at it with practice and with experience.

However, an athlete or coach who communicates well and presents with a sense of confidence, focus, and composure may be able to bring these same qualities to the locker room and to the playing field.

Also, remember that opposing players and coaches see and hear what is happening in the media, and they can be impacted by how the other side is communicating and behaving during interviews.

Since readers of this book are gathering information about teams prior to big games, the press, the media and the internet are important sources of information for you. Watch as many interviews as you can and try to determine which players, coaches and teams seem to be acting as if they are in the zone or out of the zone.

Desire And Passion

When an athlete comes to see me to improve his or her performance, I always assess his or her desire. I ask them why they play the game. I ask them what they love about the game. And I inquire about their goals and their dreams where their sport is concerned.

Obviously, some players are attracted by the huge rewards that sports can offer to outstanding athletes.

However, I believe that players who really excel and who can manage the pressure and competition which accompany a big game have a great deal of sincere passion for their sport. This kind of intense emotion can serve as a powerful motivator when the pressure is on.

Look for athletes and coaches who communicate this kind of passion and desire. These are the kinds of people who would compete in their sport even if there were little or no compensation.

Believe me, there are still players around like this today and these are the players who are likely to rise to the big occasion.

Carlton uncovered some interesting data where desire and leadership are concerned. This information should be of interest to those readers who are responsible for recruiting athletes.

Quant Fact: Competitiveness – Desire to Win and Leadership

Our research has helped to make these normally "intangible" factors (leadership, competitiveness, and "desire to win") more quantifiable. For example, if you have two players with similar skill and talent levels, the player who has "willed" his team to a championship has the higher potential to be truly great. Some of our research can be used to discover and draft players who are "under the radar" and have the potential to develop into great athletes and competitors.

Not everyone will possess this level of competitiveness. However, we believe that this can be learned or acquired. Many attributes of winners are related and can be taught and learned with practice and repetition. Underlying themes include a desire to win, focus, and control over emotions.

Quant Fact: Competitiveness and Greatness

Michael Jordan's greatness was flashed early in his career, right at the end of the 1982 NCAA Championship Game for the University of North Carolina. Just a freshman, Jordan hit the game-winning shot with seconds left in the game.

The game had been a tight throughout, with the NC Tar Heels prevailing over the Georgetown Hoyas. The game included future NBA stars like Georgetown's Patrick Ewing, and North Carolina's James Worthy, Sam Perkins, as well as the young Michael Jordan. Jordan's abilities to lead and to perform in the big game were present even at this early point in his career. ***True greatness often finds a way to win, at every level of competition they play.***

Quant Fact: "Desire to Win"

Chariots of Fire, by W.J. Weatherby begins with an interesting comment on the importance of desire and its role in competing when the contest begins.

> *"Running a race hasn't changed much in a thousand years. It is still a supreme test of the individual. When the starter's pistol cracks, runners are alone, dependent on themselves. The greater their physical and mental conditioning, the greater the runners. The immortals of the track become "chariots of fire," driven by a burning desire to win."*

"... a burning desire to win..." Everyone wants to win, but different people have different levels of that "fire inside." We try to measure and quantify this desire and the associated leadership qualities in this book – and plan to refine our ability to measure it in our future research.

Durability And Stamina

Two qualities which are somewhat related to desire are durability and stamina. Athletes and coaches with a history of hanging in there physically and psychologically can be significant assets to a team. Their experience, courage and ability to be there for the long haul can inspire and educate others as the big game approaches. Moreover, these are the kinds of warriors who are likely to show up for the big contest.

Perfectionism

Many athletes and coaches push themselves and their players very hard and are quite perfectionistic in their approach to their sport. And while this idea sounds sensible and wise, this approach backfires on some competitors.

When a person is trying to be perfect all the time, they tend to become overly self-critical. This makes it hard for the athlete to relax, to have fun, to focus and to allow his or body to do what it needs to do to play in the zone.

I encourage athletes to develop the "right level of perfection."

So, in analyzing players and teams, look for competitors who work hard, try hard but who do not turn their desire into a life or death matter.

Trying To Please Others

Some athletes are overly concerned with pleasing and impressing parents, coaches, adversaries, fans and loved ones. This kind of athlete tends to be insecure and mistakenly believes that his or her life will be better if everyone adores them for their accomplishments in sports.

Some years ago, I coached a golfer who seemed more interested in shaking hands with fans than in focusing on his game and winning the golf tournament.

Athletes who are "people pleasers" can have major emotional distractions when faced with a big game.

On the other hand, secure athletes who play for healthy reasons are apt to do better when faced with the big challenge of a championship contest.

A Casual Approach To The Big Game

People frequently ask me what great athletes do to prepare for the pressure of the big game.

Interestingly, many great warriors do not look upon the big game as being anything special. Rather, they approach it is another game or as another practice session.

They remind themselves that they have worked their craft thousands of times and it is simply a matter of doing it one more time. Furthermore, they look upon the game as a fun opportunity, not as a pressure packed life or death situation.

The fine competitors do not allow themselves to get caught up in the media hype or the hysteria which surrounds championships.

Therefore, in looking at two athletes or teams, give consideration to the team, player or coach who is calm, focused, confident, business-like and not overly excited or nervous.

Great competitors know how to contain their emotions until their mission is accomplished. This is why you see wild celebrations after a team wins the big game. It is at this point, that the participants are free to release all of their emotions. They can let go of the casualness after they have secured the victory.

Optimism-Pessimism

Not surprisingly, players and coaches who think positively and who present a sincere belief in succeeding tend to perform better when the pressure is on.

Conversely, competitors who exhibit negative thoughts and self-doubt often will not play to their fullest potential in a major contest.

Observe athletes very carefully and look for players and for organizations that communicate a strong belief in their ability and in a positive outcome for the game.

Injuries

Injuries and a history of injuries can hurt an athlete's ability to focus, remain confident and perform. An injury can be a source of distraction and cause an athlete to be tentative during the big game.

Some injured athletes can perform well in the big game probably because their body produces hormones with mask the physical pain once they start to warm up and compete in the game.

Nevertheless, you must consider the physical health of the player or the organization as they approach the big game.

History Of Choking In The Big Game

Every athlete greatest fear is choking in the big game. And if a player or a team has a history of performing poorly in a major contest, they can be haunted by this poor performance. Some athletes are traumatized by this kind or experience and some coaches are labeled as being incapable of winning the big game. Conversely, performing well in a big game and putting up some big numbers can have a very positive impact on a competitor's confidence the next time they enter the arena for the event.

Quant Fact: Choking and Championship "Factors"

The media and sports fans love to use the phrase "choking." However, choking may be the wrong way to characterize a change in performance in a big game. As was noted earlier, it is sometimes difficult for star quarterbacks or big homerun hitters to repeat their regular season performances in the playoffs or in championship games. In many cases, the numbers these stars put up during the regular season are "run up" against average or below-average competition. When you get to the playoffs, the level of play increases – so that players won't be hitting as many homers against the likes of Cliff Lee or Mariano Rivera – or scoring as many points against top defenses. In addition, the "sample size" or games in the playoffs are relatively few compared to the regular season. As a result, many players can seem to deviate from their historic averages within this smaller sample of games.

We have studied some of the key elements to winning in every major sport. Some highlights include:

- Big Game Experience
- Performing at a consistently high level of play, while minimizing errors and turnovers,
- Leadership by star players, and
- Leadership on the bench and in the locker room

Fantasies, Dreams and Daydreams

Many of the athletes I have spoken about have shared their dreams, fantasies and daydreams about their sport and themselves with me. Some have very vivid and clear dreams as to what they want and what they see themselves accomplishing in their athletic lives.

As someone once said, "Everything really worthwhile begins as a dream."

I believe that clear and vivid dreams help an athlete to realize his or her goals. That's why I encourage athletes to dream about winning the big game prior to playing in the game.

Furthermore, I believe that an active fantasy life can help people to clarify and achieve their performance goals. Players who say things like "I have dreamed about winning this game since I was a kid," are telling you how important this game is to them.

Their dream tells you a lot about their passion and their desire which I touched on earlier.

Spirituality And Religion

Some athletes have told me that they derive a lot of comfort and support from a spiritual orientation, religion, or from their relationship with God or a higher power. And many teams include spiritual leaders who lead prayers prior to a big game.

Prayer and religious rituals may play a role in improving one's performance and in helping an athlete, a coach or a team to remain hopeful, positive and inspired.

Do more religious athletes perform better? I do not know. However, spirituality and religion probably can not hurt one's performance in any way. And prayer may help to strengthen the sense of community and camaraderie.

The relationship between prayer, spirituality and performance probably would be a very interesting research project which Carlton and I may want to pursue in the future.

In evaluating competitors, you may want to give consideration to players and organizations with a spiritual or religious orientation.

Chemistry

Sometimes a quarterback and a wide receiver have a special relationship. It's as if they can read each others' mind and they seem to have the ability to enter the zone in unison.

Similarly, in basketball some point guards have a special connection with a center or a forward.

Likewise, some pitchers connect very well and perform very well with a particular catcher.

These special connections help team to enter the zone and they can help to propel teams to victories in big games. So be on the lookout for these special and magical athletic relationships.

Energy Level

Some athletes are high energy by nature and they always seem to be moving at a high rate of speed. Others appear to be more deliberate in their approach to their sport.

The key issue with regard to energy level is to find the mental and physical gears which produce the best results.

Some competitors need to be slowed down. Others need to speed up a bit.

Great competitors seem to make the game look easy even whey they are trying to win a very important contest.

Look for athletes who have the right arousal level, who conserve energy at the right time and accelerate the pace when the situation requires them to do so.

Tragedies

A death or an injury to a teammate can sometimes inspire a team to play with greater intensity. This kind of painful event can motivate players to perform very well.

Utilize the media and the internet to do your research and find out if a team is enduring this kind of hardship at the time of the big game.

Tragedies as motivators for a big game could be another interesting area for us to study in the future.

Scandals

Big time sports are no strangers to scandals and controversy. Most players and coaches can tune out these kinds of distractions. In my view, scandals have less of an impact on sports than does a tragedy. Most athletes who make it to big games have learned how to focus on the game and set aside events which are happening off the field. In fact, some find that the game is a kind of a sanctuary which insulates them from the distraction of the scandal.

It appears that personal scandals captivate the media and the public more than they impact athletes who are competing for championships.

Having said this, a new scandal involving a key player prior to a big game can be a cause for concern, depending on the player's personality and the nature of the news about the event.

Substance Abuse And Steroid Use

Athletes who are abusing drugs or alcohol or who are using steroids can create significant problems for themselves, for their coaches and their teammates.

Teams with steroid problems or substance abuse problems could be at risk for performing poorly in a big game especially if a lot of attention is being directed toward this substance abuse or steroid use.

Conclusions and Thoughts

Our results are based on observations of athletes, coaches and teams – in addition to our own quantitative research. The results are interesting and motivating because they indicate that a large portion of overall performance can be coached and taught. That is, although god-given talent and skills play a significant part in determining the top athletes in any given sport – a large portion of winning performances are clearly dependent on factors that can be studied, coached, practiced and mastered.

In addition, there are similar characteristics and qualities among top athletes and championship teams that we can measure analytically to help predict which athlete or team will do great things and/or perform at maximum potential in the big game. While there is much room for more research in some of the areas touched on in this book – and much more to learn about winning the big game – there are several common and recurring themes from our research:

- Great teams are often led by great coaches. Great coaches are able to repeat their success because there is real skill involved with motivating and leading a team. They know how to develop traits and characteristics of winning players and winning teams.
- "Big Game Experience" and the associated confidence – are persistent themes for champions, across all sports.
- To perform at a high level in big games, it is essential to minimize errors, turnovers, and mistakes. Many of these errors are related to concepts of sport psychology in nature.
- Relaxed aggression helps to achieve the proper level of intensity without sacrificing mistakes and errors.

Other Notes Of Interest, Conclusions, And Ideas For Future Research

- We plan to develop a reliable and valid mental toughness index which will include all the factors in this book as well as some additional ones. This will allow us to assess athletes and teams prior to the big game and prior to drafts.

- While not a focus of this book, we believe that genetics, biology, biochemistry, and physiological psychology will help us to understand more about the human mind, the human body and performance in the big game in the future.

- Our data indicate that interpersonal issues like coaching and relationships have a lot to do with winning. This is encouraging, since we believe that coaching skills and interpersonal relationships can be improved and strengthened through counseling, training and education.

- Top athletes benefit from mental training so that they can learn how to move their mind and their bodies into the zone more often. Mental training can give athletes, teams and coaches the edge they need in the important contests at every level of play.

- Coaches would be wise to learn and model what the great coaches have done to perform well in big games.

- Relationships among players, coaches and management are complicated. Improving on these vital elements is likely to result in better performances in important contests.

- Coaches, athletes, teams, managers, athletic directors, owners and sports executives can all improve their performances if they are aware of the wide range of mental factors and sports statistics which impact the outcomes of big games.

Scorecards:
Two Ways To Analyze The Big Game

We put together two scorecards to help you predict the outcomes of championship sporting events. As we explained at the start of this guide, some variables are easier to quantify than are others. Some of these factors will be the subject of future studies we plan to do, as we continue to work with teams and athletes.

Scorecard I includes all the major variables noted in this book.

Scorecard II is a "short-list" that includes the variables in the main categories which have the most quantitative and statistical support.

Feel free to integrate these scorecards with the data and models you are currently using. And feel free to reach out to us at anytime.

Who Will Win the Big Game?
Scorecard I: Complete List

Factor	Team A	Team B
1. Confidence		
2. Big Game Experience		
3. Coaching		
4. Easy to Coach		
5. Accepting Responsibility		
6. Respecting Adversary		
7. Player-Coach Relationship		
8. Momentum		
9. Margin of Victory Last 5 Games		
10. Errors		
11. Errors and Momentum		
12. Winning Ugly / Ability to Win		
13. Quality of Opponents		
14. Strategy		
15. Genetics & Mother Nature		
16. Speed, Quickness		
17. Knowing How To Relax		
18. Relaxation, Anxiety & Confidence		
19. Relaxed Aggression		
20. Focus		
21. Managing Big Moments		
22. Distraction Control		
23. Playing in the Present		
24. Visualization and the Zone		

25. Fun		
26. Resiliency		
27. Managing Emotions		
28. Anger and Frustration		
29. Consistency		
30. Hard Work		
31. Practice		
32. Home / Away		
33. Goals and Goal Setting		
34. The Locker Room		
35. Stars as Leaders		
36. Number of Stars		
37. Team Chemistry		
38. Family Relationships		
39. Body Language		
40. Demeanor and Routines		
41. Managing the Media		
42. Desire & Passion		
43. Stamina & Durability		
44. Casual Approach to Big Game		
45. Optimism – Pessimism		
46. Injuries		
47. History of Choking		
48. Spirituality and Religion		
49. Energy Level		
50. Other Life Events		
Overall		

Who Will Win the Big Game?
Scorecard II: Short-List

Factor	Team A	Team B
Big Game Experience		
Coaching Leadership		
Leadership on Field		
Momentum		
Error-Free Play		
Quality of Opponents		
Consistency		
Motivation / Focus		
Team Chemistry		
Other		
Overall		

Dr. Granat's Books And Programs

Dr. Granat has written a number of books and developed a number of peak performance programs and mental toughness programs:

How To Get Into Zone And Stay In The Zone With Sport Psychology And Self: Hypnosis. Two CD program used by top athletes from every sport from around the world. Get it at www.StayInTheZone.com

Bedtime Stories For Young Athletes. Motivate your kids and teach them how to stay relaxed, confident and focused and they enjoy their favorite sport. Two CD program. Plus a free book for parents. Get it for your child now at www.StayInTheZone.com.

How To Lower Your Golf Score With Sport Psychology And Self-Hypnosis. Two CD Program. Get it at www.StayInTheZone.com

101 Ways To Break A Hitting Slump With Sport Psychology and Self-Hypnosis. Three CD Program and a free book. You can get this program at www.StayInTheZone.com

How To Bowl In The Zone With Sport Psychology And Self-Hypnosis. Add 29 Pins To Your Score. Two CD Program. Get it at www.StayInTheZone.com

Get Into The Zone In Just One Minute: 21 Simple Techniques To Improve Your Performance. This books shows simple ways to improve your performance quickly. It is available at www.amazon.com and www.barnesandnoble.com.

Nervous about the big exam? Conquer Test Anxiety Hypnosis, Visualization and Innovative Thinking. A 5 CD program available at www.ConquerTestAnxiety.com

21 Creative Ways to Conquer Stress & 18 Simple Ways to Relax with Self-Hypnosis, Meditation & Visualization. Three CD program to help you handle every day stress. Get it at www.StayInTheZone.com

Zone Tennis. Master the mental game of tennis with simple tips. Play your best tennis now. Get your copy of this book at www.Amazon.com

Want Your Athlete Or Team
To Perform Better?

The authors of this book can help you to understand what you need to do to have your team win more often and perform to its fullest potential.

Carlton Chin, CFA, will show you where you are weak and where you are strong by doing a thorough inventory and statistical analysis of your team and your competition. These methods can also be applied to improve the results of scouting reports and drafts – as well as to analyze potential trades.

Dr. Jay Granat will show you and your athletes what you need to do to get mentally tough and to get into the zone more often.

To find out more about these comprehensive services, contact us at info@stayinthezone.com or at 888 580-ZONE.

About The Authors

Jay P. Granat, Ph.D

Jay P. Granat, Ph.D., is a Psychotherapist, hypnotherapist and a licensed marriage and family counsellor. The founder of www.StayInTheZone.com, Dr. Granat has coached athletes and their families from virtually every sport from around the world. He has worked with an Olympic gold medallist, Olympic athletes, professional golfers, tennis pros and elite young athletes from many sports and has lectured to many teams, sports organizations, clinics, clubs, camps and to some of America's largest corporations.

A former university professor, he writes a weekly column for four newspapers and has appeared in The New York Times, Good Morning America, Forbes.com, The BBC, The CBC,

Sporting News, ESPN, The Newark Star Ledger, ESPN, Tennis Magazine, Tennis View Magazine, Iowa Golfer, Executive Golfer, New York Family Sports and The Bergen Record. Golf Digest named Granat one of America's top ten mental gurus.

Jay Granat, earned his Masters and Ph.D. in Counseling from The University of Michigan. He is the author of To Get In The Zone And Stay In The Zone With Sport Psychology And Self-Hypnosis, How To Lower Your Golf Score With Sport Psychology And Self-Hypnosis, How To Conquer Test Anxiety and How To Bowl In The Zone With Sport Psychology And Self-Hypnosis and How To Break A Hitting Slump With Sport Psychology and Self-Hypnosis, 21 Creative Ways To Conquer Stress and 18 Ways To Relax With Self-Hypnosis, Meditation And Visualization How To Get Into The Zone In Just One Minute and Zone Tennis.

Granat is particularly interested in the role that family relationships play in stress and in one's ability to perform their best when under pressure. He is past Vice President of The New York Society For Ericksonian Psychotherapy And Hypnosis a member of The American Psychological Association and The American Counseling Association.

Dr. Granat recently founded www.TopSportsDoctors.com and www.USASportsDoctors.com. These sites will serve as referral network to match up athletes, coaches and families with professionals, coaches, equipment and resources which can help them to perform their best.

Granat welcomes your questions and can be reached at 888 580-ZONE or at info@stayinthezone.com

Carlton J. Chin, CFA

Carlton J. Chin, CFA, is principal and founder of CARAT Juroca LLC, a quantitative research consultancy and investment manager. At MIT, Carlton developed a baseball simulation program to study baseball sabermetrics and the impact of varying batting orders. Carlton has applied his extensive knowledge of mathematics and statistics to a wide range of sports.

An MIT-trained "quant," Carlton has used mathematics and numerical simulation in fields ranging from the financial markets to mechanical engineering. He has worked as an actuary and was selected as a proprietary trader for Goldman Sachs/Commodities Corp. He currently manages several "alternative investment strategy" funds in addition to his work as a research consultant.

Carlton completed the requirements for his SM and SB degrees at MIT at the age of 21. He is a CFA (Chartered Financial Analyst) and has been featured and quoted by various financial industry publications. He serves as an MIT Educational Counselor and is a member of the AIMR, NYSSA, and USTA.

Carlton is particularly interested in alternative investment strategies and the futures markets – where he believes investment managers can add the most value. He also enjoys applying probability and statistics – to fun and interesting projects – to maximize performance and value, given restrictive parameters.

For more information, please view Carlton's information at the links below. He welcomes your questions, input, and inquiries.

http://www.linkedin.com/in/carltonchin

http://caratcapital.com/

A Bit More About The Authors

Carlton and I have been pals and tennis partners for more than twenty years. We have spent dozens of hours discussing sports, sport psychology, and statistics related to sports and to life. We often talk about who will win a game or a contest. Carlton highlights the data, the numbers, and the probable winner based on math. He is a real math whiz and can apply probability and statistics to answer a variety of questions and solve a range of problems. In this book, he applies his fascination and expertise with numbers to the analysis of the big game in several different sporting events.

I talk about things like leadership, motivation, mental toughness, focus, relationships, the zone and peak performance based on my counseling work with athletes, coaches and families.

This guide combines my ideas and Carlton's ideas. That is, this book blends math and psychology to evaluate athletes and sporting events in order to determine which competitor is best – and identify the likely winner of the big game.

Thanks

Carlton and I have known each other for quite some time. We share a number of interests and passions. We have different but highly complimentary ways of looking at the world and of solving problems.

Collaborating on this book with such a good friend has been a real joy for me.

Carlton is super smart, hard working and very efficient. He always delivers on what he says he will. He is the kind of person you can count on to get things done.

More importantly, he is also a thoughtful and considerate gentleman. It has been my privilege and my pleasure to work with him on this book.

We both have learned something about teamwork from this project and our friendship has grown through the process. We hope our readers enjoy this guide, since we are already talking about the next book.

Carlton, thanks for all your help, your energy, your smarts and your enthusiasm.

I also want to thank my lovely wife, Robin. She always supports my writing and me with optimism, encouragement and an incredible spirit.

Last, I want to thank Mike Strozier, my publisher, for always responding positively to my manuscripts. This is my third book with World Audience, Inc.

Jay Granat

—Get Into The Zone In Just One Minute: 21 Simple Techniques To Improve Your Performance ISBN: 978-1-934209-63-9

—Zone Tennis by Jay P. Granat, Ph.D. ISBN: 978-0-9820540-9-3